Small Format
(and Mixed Format)
Interchangeable
Core
Servicing

By Don OShall

Locksmithing Education Publishing, LLC
First Edition, © January 2014,
Second Edition, © January 2017
by Donald O'Shall,
Beverly Hills, FL - USA

ISBN 978-1-937067-05-2

Library of Congress Cataloguing-in-Publication Data

OShall, Donald A

Small Format (and Mixed Format) Interchangeable Core Servicing
First Edition / by Donald A. O'Shall

ISBN 978-1-937067-05-2
1. Industry—Security Measures—Management

ISBN 978-1-937067-05-2 $45.00 USD

54500 >

9 781937 067052

Locksmithing Education Publishing, LLC
12 South Lee Street
Beverly Hills, FL - 34465

http://www.Locksmithing Education.com

rev 2017-01.1

Table of Contents

This book is dedicated to

Bill and Jill Young,

without whom it would probably never been written, and to

*Ron Marcinkowski
and
Bill Neff,*

*who started it all
with a phone call in 1984,*

and to my friend

Terry Lantz,

*who helped me with the first draft
on an old manual typewriter in 1984.*

Chapter One:
History and Development of the Core

In 1844 Linus Yale, Sr., a noted inventor who specialized in bank vault mechanisms, invented the pin tumbler lock. The key end was a round cylinder with bittings milled into it from four directions.

In 1853 he was granted another patent, number 10,144, for another pin tumbler lock. Progress had been made and coil springs were now in use in the design.

In 1857 he obtained yet another patent, (Number 18,169), for a pin tumbler lock. This one has become known as the Yale Slide padlock because of the sliding action of the two sides once the correct key is inserted.

Texts on the history of locks often point to the early (approximately 2000 bc) Egyptian (or Phoenician) lock as the first example of a pin tumbler lock, but even a cursory examination of this lock will tell the locksmith that it really is the fore-runner of the disc lock (or wafer lock) rather than the pin tumbler lock.

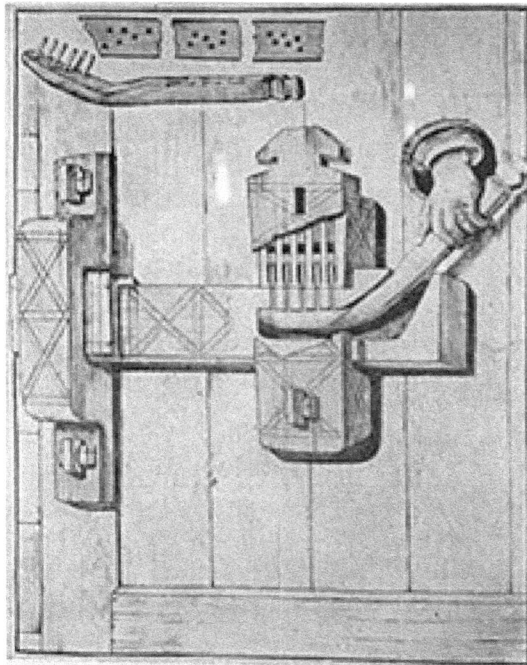

Example of the ancient Egyptian Lock

The first metal lock appeared between 700 bc and 900 bc.

Yale's son, Linus Yale, Jr., improved on his father's patented design and in 1861 and 1865 patented his improved versions of the pin tumbler lock, then eventually improved on those by using flat keys instead of milled bit keys, and then by using serrated keys instead in the slide padlock design.

In 1898 Yale's company patented the paracentric key (designed by Warren Taylor), which unlike the serrated key, had a pattern of grooves that crossed the centerline (para- for near or beyond and –centric for referring to the centerline) making it even more secure.

With the flat key, any thinner blank was capable of operating the lock if properly cut. The serrated key's "wobbly" shape made this far more difficult but was still a possibility. With the paracentric key, the grooves crossing the centerline prevented this. It had to have the same pattern in order to enter and operate the lock.

He also invented the mortise cylinder, the first cylinder to be removable from the lock body, which although it is not considered a removable core by today's definitions, was a vast improvement in both convenience and security, and no doubt helped inspire the removable core concept.

L. YALE, Jr., Dec'd, S. N. BROOKS, Adm'r.
Assignor to YALE LOCK MANUFACTURING CO.
Lock.
No. 8,158.   Reissued April 2, 1878.

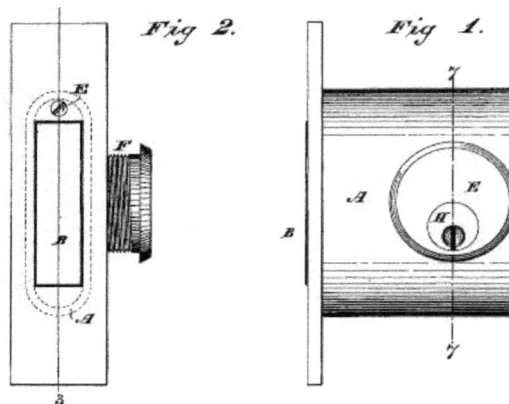

Example of a modern mortise cylinder and Yale's patent drawing of the original one.

According to Tom Hennessy, (curator of the Lock Museum of America in Terryville, Connecticut since its foundation in 1972 and developer of the Standard Key Coding System in popular usage today), Master keying of Lever Locks can be traced back to as early as 1686, and Master keying of Pin Tumbler Locks appeared in an 1865 Yale & Greenleaf advertisement.

Noted Master Keying expert and author Billy B. Edwards further notes that it can be determined from very old Yale catalogs that the first master key systems, (those offered by Yale up until 1872), were in fact shoebox (no design or pattern of controls built into the system) key systems.

"That was evident by the catalog claim that an unlimited number of Keyed Different locks could also be operated by a single MK. But by 1878 Yale keying personnel had discovered the principles of the RCM (Rotating Constant Method) and routinely implemented four constants in their five pin cylinders. That fact was evident by the catalog claim that you could have 1 MK with up to 20 Change (individual operating) Keys.

There is in fact an engineering drawing labeled #4557E and 4457E dated 9/11/1896 that indicates they discovered the TPP (Total Position Progression) that year and that the TPP was used for the Surety System. "

However, these systems had a disadvantage in that "split pin" master keying (the standard methods applied to pin tumbler locks) lowered the physical integrity of the lock cylinder and made it more vulnerable to attacks such as picking, key manipulation and system deconstruction.

The Master Ring cylinder defeated most of the problems of split pin master keying. It was patented by Edward O'Keefe, a New York City locksmith, in 1889 and subsequently purchased by P&F Corbin Company. Later it was produced in both the Corbin and Russwin keyways.

This cylinder uses the rather unique principle of a plug within a ring which acts as a second and independent plug. This was the first dual shearline product, and was the predecessor of the interchangeable core.

The theory of the concept is that the CHANGE KEY (individual operating key) is lined up at one shearline and the master key is lined up at the other shearline.

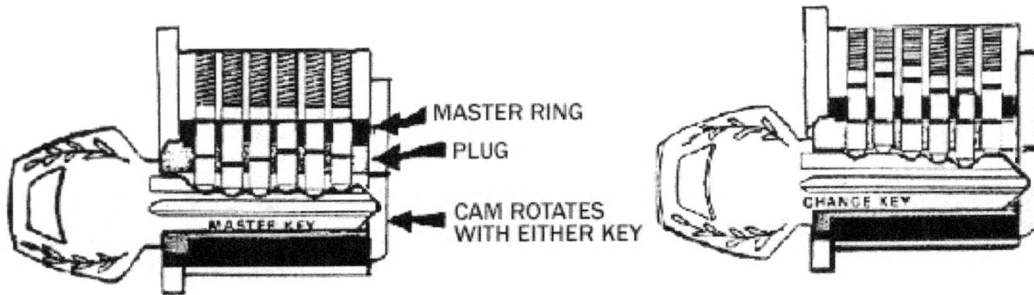

The Master Ring Cylinder Operation

In the illustration on the left, the driver pins (top pins) are all lined up above the upper (Master Ring) shearline, while a combination of Bottom Pins (the conical bottomed pins intended to come in contact with the key) and the Build-up (a new type of pin invented for this cylinder) pin fills the space between the bottom pin and the driver pin.

In the illustration on the right, the bottom pins are all lined up at the bottom (operating) shearline, while the upper (master ring) shear line is blocked by either the drivers or the build-up pins in each position.

If even a single position is blocked at the active shearline, the key cannot turn at that shearline.

In 1923, Frank E. Best invented a new type of padlock. He continued to make modifications to it and was granted new patents in 1924 and 1926. The following illustrations show the 1923 patent drawing and the 1924 and 1926 modifications.

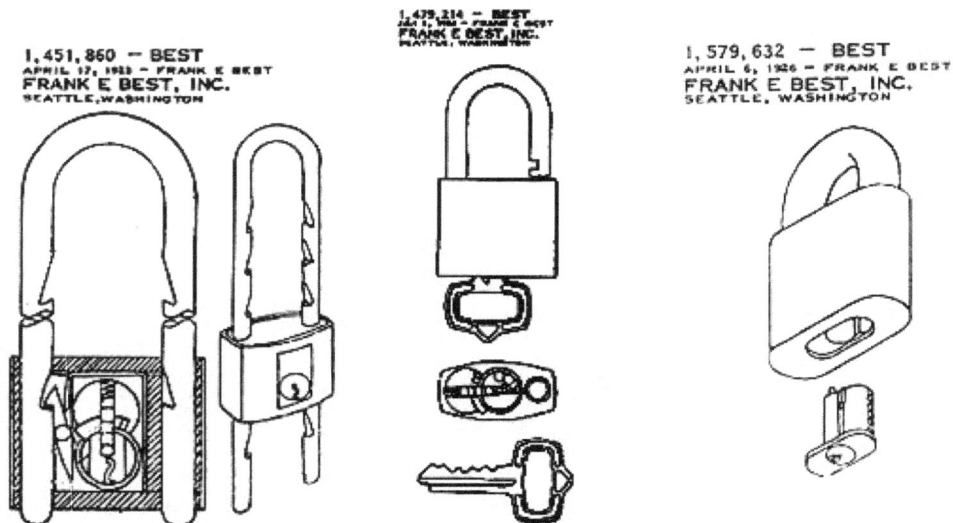

The original Best lock patent drawings.

These locks used a removable core which contains the bottom pins, master pins (if any), the drivers and springs and an additional pin between the master pins and drivers that can be referred to as a build-up pin or a control pin.

The cores of the lock made use of a second shearline, similar to that of the Master Ring.

Unlike the Master Ring Cylinder, however, the second shearline was not round the whole way around the plug, but instead was a "control sleeve" just wide enough to create a "control lug" that is capable of securing the core into the padlock. The difference can be seen in Fig. 1-6.

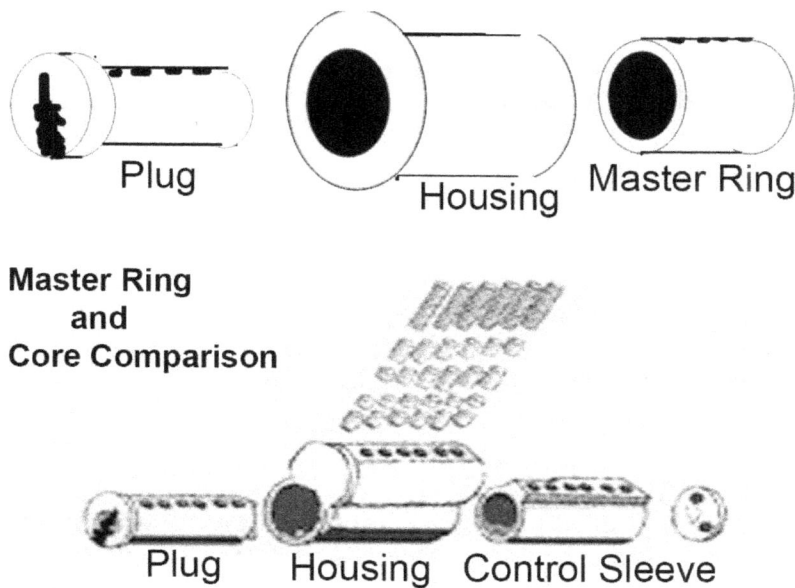

Plug   Housing   Master Ring

**Master Ring
and
Core Comparison**

Plug   Housing   Control Sleeve

Fig. 1-6) Parts of a Master Ring Cylinder and Parts of a Best (SFIC) core

Best continued to expand and improve its new product, and began a continuous effort to apply the new core to all types of locksets through the use of specially designed "housings", eventually licensing other companies' patented locksets to apply it to, and convincing them to license Best to make adaptor knobs to fit their products.

Because it was able to fit into all their housings, Best called it an "Interchangeable Core".
In the beginning Best was the only manufacturer of these types of cores. However, not understanding their market fully, they only sold through Best primary distributors and would not sell their products to locksmiths or wholesalers, only to the end users and any institutional locksmiths they may have.

That might seem quite logical, and probably reduced costs for the end user in theory. However, the vast majority of institutions at the time did not have institutional locksmiths working for them, and if they did they were usually under-trained, under-skilled and underpaid.

Therefore the mainstream locksmith was frequently called to service these accounts. But he could not stock the parts and locks he needed unless the end user was willing to pay for it and order it, and not all end users understood the importance of doing so.

This created a gap in the market. Best tried to correct for it by hiring their own factory locksmiths to service these smaller accounts, and by providing the first factory training programs for institutional locksmiths, but the demands of institutions tended to be more time sensitive than such a system allows.

However, by 1967 Best still did not offer a heavy duty knob lock. Falcon Lock, recognizing the potential for a heavy duty locket in the Institutional and industrial markets that Best was dominating, created a heavy duty lock designed to take the Best core. Unfortunately for Falcon, Best declined to partner on it. In fact, Best declined the offer and instead continued to persuade other manufacturers to make adaptor knobs for their products.

This left Falcon hanging with a new product with no core. As a result, Falcon decided to design a core that defeated any outstanding patents (most had already expired) and sold it with their new lock, creating the Falcon Interchangeable core. This was a product almost identical to the Best core except that it used a slide cap to cover all the pins and springs instead of the individual caps used by Best.

The Falcon interchangeable core fit into the Best housings as well as any other locksets intended to accept the Best Housing. Its initial differences were manufacturing materials, a brass "spring cap: pressed into place instead of individual chamber caps and a C-clip to hold the plug in the core instead of a pressed on retainer. (Today, Falcon Interchangeable Cores are available with either configuration, as are many other cores similar in design.)

Best responded to this by first licensing, and then producing, a heavy duty lockset.

Falcon dominated the "Best type" core market among mainstream locksmiths for many years, although a few other products were introduced as well during that time, many with unique features intended to be improvements.

Other manufacturers took a different approach. As all this was happening, other manufacturers had been making "Removable Cores" such as the early Yale R-core which was updated to a more modern type of core in 1959. Yale, Sargent, Schlage and several others used the principle of the single position sleeve, with various solutions but all operating in a single position.

Some later realized that the more positions the control sleeve covers, the better the physical integrity of the core itself, and the less likely someone could recreate the control key, and expanded to a newer control mechanism covering two or more positions.

From 1964 to 1971, the Corbin and Russwin Lock companies made a rim and mortise housing that accepted a round full cylinder with a round control sleeve (much like a master ring) that operated off the single position at the tip of the key. This was not considered interchangeable because it did not fit any of the knob locks that either one of the manufacturers offered. They were referred to as "Removable Cores".

In 1971, they introduced their more versatile "removable" core, which used a control sleeve that basically was a saddle sitting on the four "middle" positions of a six pin plug or their equivalent positions on a five pin plug. In 1985, they updated it to a control sleeve wrapping around the plug much as the Best control sleeve did, which made the core easier to handle during rekeying and other service procedures than their earlier cores. But the control sleeve, like its saddle-type predecessor, existed in only four positions, with standard cylinder chambers in the remaining positions.

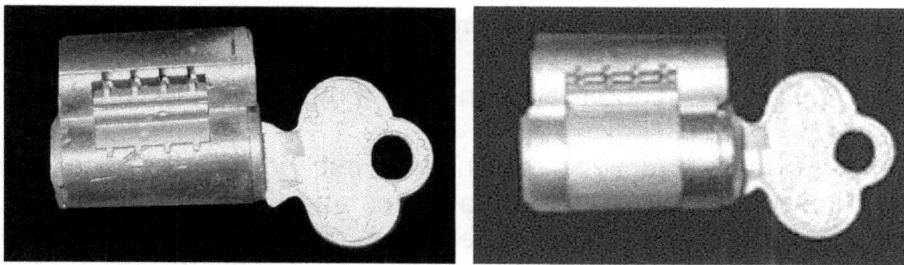

One of the biggest differences between the Corbin and Russwin cores and the Best and Falcon Cores is that the Best and Falcon required a special pinning kit. Their pin tumblers are approximately .110" in diameter, while the standard lock cylinders use .115" diameter pin tumblers, which is larger, and resulted in the need for a slightly larger core. Eventually, products like this would become known as Large Format Interchangeable Cores, or LFIC products.

The other difference from Small Format Cores is that, as mentioned earlier, the Corbin and Russwin control sleeve does not cover all the pin chamber positions, as the Best and Falcon and similar products do.

As Best patents expired and it became clearly legal to do so, other manufacturers began to make their own versions of the "Best type core". Other manufacturers jumped on board before long with identical sized products that fit the same housings, some with design improvements that even improved the security of the core itself.

Locksmiths referred to any core that was not "Best-type" as a "Removable Core", arguably because it was not interchangeable between manufacturers as the "Best-type" interchangeable cores were. Most manufacturers followed suit in their own literature, but not all did, which led to some degree of confusion.

"Best-type" was a term which none of the manufacturers other than Best liked particularly much.

All of the "Best-type" cores were intended to fit into the hole in the cylinder housing as originally designed by Best. But can you imagine having your improved product, perhaps even a high security product, described in terms of a competing manufacturer? Schlage lock was the first to do something about it.

They had initially created their own product, similar to that of Yale and others, using full size pins and standard cylinder keys, with a control sleeve in a single position beyond the tip of the standard key, and requiring a specially designed key to remove the core. They began calling it a "Full Size Interchangeable Core" or FSIC.

When Schlage decided to add an interchangeable core to their line-up, they decided it was time to stop using the term "Best-type". At this time Schlage was employing noted lock authority A.J. Hoffman, CML, who is credited with coining the terms "Small Format Interchangeable Core" (or "SFIC") in place of "Best-type", and "Full Size Interchangeable Core" to describe their standard "removable core" product.

The term "Small Format" had two references. First, some other manufacturers had made cores and housings that used standard pins and therefore required a larger hole to put the core in. Secondly, the "Best type" cores cannot be pinned out of a standard locksmith pinning kit, because the diameter of the pins is slightly smaller (.110" instead of .115") for the "Best Type".

Smaller hole cutout in the housing and smaller diameter pins equals "Small Format". Best called its cores "interchangeable" because you could use it in all their product lines. Later the term came to be associated with the fact that it was also interchangeable among various manufacturers' products which were intended for their own "Best Type" cores.

Locksmiths and other manufacturers immediately jumped on the term "SFIC" or "Small Format Interchangeable Core", but the term FSIC did not catch on, other than for the Schlage large pin product. It was instead replaced by the term "Large Format Interchangeable Core" or "LFIC", which they also applied to the Schlage FSIC.

Today, the term "Removable Core" is usually limited to specially designed cores used primarily by some padlock manufacturers, although the adjective "removable" is still often used in describing interchangeable cores.

Interchangeable cores of both the Small Format and Large Format type continue to be popular today, and in fact seem to be constantly increasing in popularity. The ability to respond quickly to rekey a door whose key has been compromised is vital to the security of most institutions.

They do not fit as well into residential security as well, however, except for apartment buildings. The average homeowner is simply not equipped to track a control key and keep spare cores lying around the house securely, and granting a locksmith responsibility for the control key (the key to remove and change cores) did neither party any favors.

Residential lock manufacturers continue to try to find alternatives such as the smart keys and smart locks that were introduced in the early years of the 21st century, and the biometric locks. So far, these have not been a whopping success and many are discarded nearly as quickly as they are introduced.

Meanwhile, in the institutional market, and especially the apartment complex market, the interchangeable core offerings continue to grow and improve in security. Apartment owners, realizing the risks caused by the ability of a homeowner to disassemble the lock cylinders without anyone noticing, and who have experienced a rise in thefts by master key, are frequently requesting cores instead. Interchangeable cores typically require the control key in order to access the core for disassembly, offering a large increase in security.

Also, many manufacturers have introduced restricted keyway products, patent protected keyway products, and products with secondary locking mechanisms such as finger pins or sidebars to improve the interchangeable core even further.

About Mixed Format Cores

One of the bigger differences between most of the Small Format Interchangeable Core products and most of the Large Format products is the fact that the Small Format products usually have a control sleeve that covers all of the chambers, while the Large Format rarely do.

That means that when calculating the pinning for a Large Format core, you have to bear in mind which chambers get the pinning calculations for the control pins, because the other positions, not having a control sleeve to compensate for, do not.

Corbin-Russwin standard cores have a control sleeve in four positions, which can be a bit confusing if you also are using a five pin key in a six pin capable core! (Even more confusing can be the fact that some of their keys are numbered bow to tip and others tip to bow.) Corbin-Russwin Interlocking Pin (formerly Emhart) used two positions only, and while the product used interlocking pins, these positions did not.

Sargent originally had two different models of its Large Format cores. The older one (series 5100) had only one control sleeve position, at the tip, and the newer ones use two positions in chambers three and four in the middle of the core. Today they have added a higher security product to the line as well.

These are just some examples. But the important thing is that you realize that the pinning for non-control sleeve positions will be different from the pinning for control sleeve positions, and that this varies among Large Format manufacturers.

Further, with the control sleeve affecting fewer positions, accidental core removal when inserting or removing keys is a common complaint. The less positions required, the more likely the problem becomes. Preventing this was the primary reason Schlage designed its control key to be longer than operating keys in the FSIC product lines.

To solve this and permit the continued use of the typically expensive hardware involved, some manufacturers of Small Format Interchangeable Cores accomplished a useful modification. They created cores that use the Small Format pinning but fit the overall outside shape of the Large Format Cores of other manufacturers.

In other words, the pin diameter will be .110" and the pinning calculations are accomplished as though you were pinning a Small Format Core, but the actual pin lengths may (but do not always) vary from those in a standard pinning kit for SFIC. the cores look like, and fit into, each manufacturer's LFIC core housings.

The first of these was Peaks. Peaks had originally made a higher security "Best-type" core. It used a secondary locking mechanism in the core which sat on a "peak" at the shoulder of the key. Using Patent Protection, it offered a degree of Key Control that was new to the industry.

It also was the first to create and market Mixed Format Interchangeable Cores (MFIC). Using the same pinning and the same small diameter of pin, it lengthened the bottom pins to fit a larger plug diameter, and made the outside shape that of other LFIC products. But the beautiful part is that it had a control sleeve covering all the positions.

Medeco followed the same sort of pattern once it began producing the Arrow cores. Although their standard product was an LFIC, Medeco also produced several products of Mixed Format. It also took the process a step farther in that it offered multiple products of varying degrees of protection. Some featured a sidebar. But not all of their Mixed Format line were clearly definable. They also produced an interlocking pin model to fit other manufacturer's LFIC housings. So, some care must be taken to determine exactly which product you have before attempting servicing.

Overview of SFIC and MFIC

Best originally numbered their uncombinated (no pins) core A1, and their first pinned cores A2. When they changed the numbering system for a different master key system using different key depths for government and hospital usage, they called the new one A3. Then they made another similar change for Universities and large complexes, which they called A4. Other manufacturers adopted the original Best numbering systems for the three pinning calculations.

Best keys use a tip stop and have no shoulder. A seven pin key can only operate the five pins of a five pin core one way, not create other combinations as it is run in and out of the key. Key numbering is listed from tip to bow, and cores are normally held for pinning with the keyway (face of the core) to the right.

This is not necessarily true of all manufacturers making what seem like similar products. Once again, not every manufacturer of similar products does this, so use care when pinning to other manufacturer's keys for Interchangeable Cores or you could end up with cores that do not work.

Best numbered the A2 Bottom Pins with an A and the A2 Top Pins with a B.
Best numbered the A3 Bottom Pins with a C and the A3 Top Pins with a D.
Best numbered the A4 Bottom Pins with an E and the A4 Top Pins with a F.

All manufacturers of SFIC use these pin designations. However, for Mixed Format Interchangeable Core (MFIC), other letters are typically used when the pin lengths are changed to fit a larger plug diameter.

J,B (A2 Mixed Format to fit Kaba for Corbin-Russwin Removable Core)
K,F (A4 Mixed Format to fit Kaba for Sargent Removable Core)
      and Kaba 3400 large diameter (-1095,-1099)- (various-with screw cap)
X,W (A2 Mixed Format to fit Kaba for Schlage Full Size (FSIC)

The A2 numbering system is .0125 increments from 0 to 9 with 9 deep.
The A3 numbering system is .0175 increments from 0 to 6 with 6 deep.
The A4 numbering system is .0205 increments from 0 to 5 with 5 deep.

A zero cut on all three numbering systems is the same depth.
Zero Bottom Pins are therefore called 0ACE.

A zero cut leaves approximately .3175" on the key,
and a zero Bottom Pin is .110" to match it.

The effective plug diameter for SFIC is approximately .428"; Mixed Format (MFIC) vary depending on manufacturer and intended application.

The Control Sleeve is generally listed as .125";
which is the measurement most tests consider to be the correct answer,
although measurement is more likely to come in at closer to .123";

It should be noted that there is no difference physically between an A2, A3 or A4 core until it is pinned. All three have been available in five pin, six pin and seven pin length, although it is exceedingly rare to see an A3 or A4 system that is not 7 pin (seven cuts per key).

Pin Sizes (approximate):

Depths		A2 BP	Depths	A2 TP	Depths	A3 BP	Depths	A3 TP	Depths		A4 BP	Depths	A4 TP
0A		.110"			0C	.110"			0E		.110"		
1A		.122"			1C	.128"	1D	.018"	1E		.131"	1F	.021"
2A		.135"	2B	.025"	2C	.146"	2D	.036"	2E		.152"	2F	.042"
3A		.147"	3B	.037"	3C	.164"	3D	.054"	3E	0K	.173"	3F	.063"
4A		.160"	4B	.050"	4C	.182"	4D	.072"	4E	1K	.194"	4F	.084"
5A	0J	.172"	5B	.062"	5C	.200"	5D	.090"	5E	2K	.215"	5F	.105"
6A	1J	.185"	6B	.075"	6C	.218"	6D	.108"		3K	.236"	6F	.126"
7A	2J	.197"	7B	.087"			7D	.126"		4K	.247"	7F	.147"
8A	3J	.210"	8B	.100"			8D	.144"		5K	.268"	8F	.168"
9A	4J	.222"	9B	.112"			9D	.162"				9F	.189"
	5J	.235"	10B	.125"			10D	.180"				10F	.210"
	6J	.247"	11B	.137"			11D	.198"				11F	.231"
	7J	260"	12B	.150"			12D	.216"				12F	252
	8 J	.272".	13B	.162"			13D	.234"					
	9J	.285"	14B	.175"									
			15B	.187"									
			16B	.200"									
			17B	.212"									
			18B	.225"									
			19B	.237"									

Spool bottom pins to resist picking are available in 7A/2J, 8A/3J,and 9A/4J for the A2 system. The appropriate number is 7AS, 2JS, 8AS, 3JS, 9AS, 4JS respectively.

Spool top pins to resist picking are available in 6BS, 8BS,and 10BS for the A2 system. Spool bottom pins to resist picking are available in 4E/1K, 5E/2K for the A4 system. The appropriate number is 4ES, 1KS, 5ES, 2KS respectively.

Spool top pins to resist picking are available in 4FS, 5FS,and 6FS for the A4 system.

Let's examine each series:

Chapter Two:

The A2 Small Format Interchangeable Core

These locks use a removable core which contains the bottom pins, master pins (if any), the drivers and springs and an additional pin between the master pins and drivers that can be referred to as a build-up pin or a control pin. They operate on a principle similar to the Corbin-Russwin master ring cylinder, except that on these the ring only goes part of the way around the plug, and serves the function of holding the core in the housing. Also, because the upper shearline is used to operate the control lug which holds the core in place or releases it, it only can be pinned one way, making it simpler.

The beauty of the principle behind this lock cylinder is that lining some of the pins up at the standard shearline and some of the pins up at the upper control shearline accomplishes NOTHING! All of the pins must be lined up at one shearline or the other in order to turn the key. Because of the design of the control sleeve, it can only rotate a quarter turn and cannot actually operate the lock.

The upper shearline is used to move the portion of the ring that is used as a control lug to hold the core in the cylinder, thus releasing the core from its housing when turned one eighth of a turn. Figure 2-1 shows the parts that make up a typical small format interchangeable core.

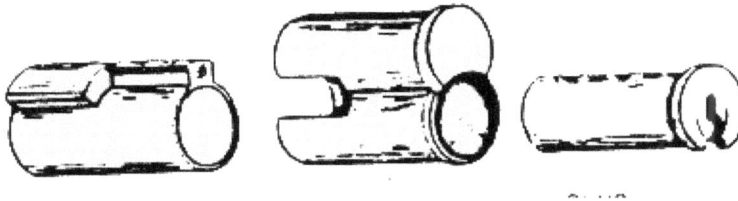

Fig 2-1) Parts of an SFIC core

The control sleeve is rotated when a set of pins blocks the lower (standard) shearline, and lines up at the upper (Control Sleeve) shearline. These pins are generally called a BUILD-UP PIN in most other types of locks, but are referred to as a CONTROL PIN in the Small Format (and mixed format) Interchangeable Core products. It makes up for the difference between the Bottom Pin (and the Master Pin if it exists) and the Control Sleeve shearline when the key is inserted that should operate at the upper (control sleeve) shearline.

To understand this better, we will use an example of just one chamber and no extra master pins (split pin method) needed.

If we have a three cut on both the master and individual key in a specific position, the individual key will bring the number three Bottom Pin to the operating shearline.

But if we had another key with the same cut in that position that we wanted to operate at the upper shearline as well, it would bring the Bottom Pin to the standard shearline, but we would still need to make up the distance from there to the control sleeve shear line.

That means the build-up pin (control pin) has to be the exact thickness of the control sleeve because there is no difference between the two keys. The thickness of the control sleeve is approximately .125 inches (one eighth of an inch). Therefore, in order for our second key to line up pins at the upper shearline as well, it would need a control pin with a length of .125".

The key system referred to as the A2 system uses depths of 0 to 9 with increments of approximately .0125 (twelve and a half thousandths of an inch). The .125 is approximately 10 of the A2 increments in thickness.

(Actually the increments are typically about .1225" and the control sleeve is typically about .123" but this does not match published information or test questions.)

Fig 2-2) The 10 Control Pin is the thickness of the control sleeve.

Now we need to know what to call such a pin. Well, because it is the length of control pin that will be used whenever there is no difference between the two keys, and is equal to the thickness of the control sleeve which is equal to 10 increments, why not call it a 10 control pin? That is exactly what the manufacturer decided as well.

But what if they were not the same? This is quite likely, because if every chamber had a 10 control pin needed, there would be no way to make it operate one shearline or the other. Both shearlines would be free spinning when the key was inserted, which would be an undesirable outcome.

So it is likely that few of the chambers would actually use the same cuts on both the operating key and the control key.

What if the cut on the second key had been two increments deeper, in this example a five cut?

Now instead of bringing the three bottom pin to the standard shearline, it brings it two increments too short for the standard shearline, which allows the control pin to drop into the plug by two increments, effectively blocking that shearline.

That is actually a good thing, because we WANT our two keys to operate at different shearlines, not open both. But it also means that the 10 control pin will not reach the control sleeve shear line either. It is too short by two increments, so instead of a 10 we need one that is two increments longer.

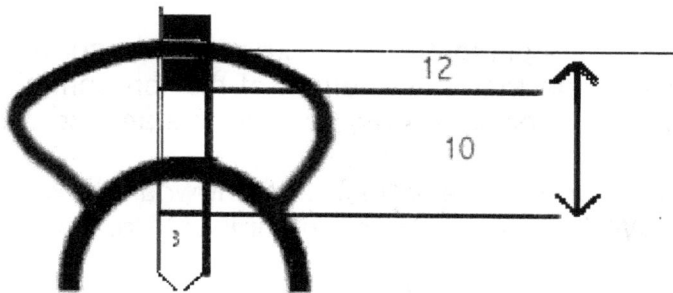

Fig 2-3)   The 12 Control Pin makes up for the 2 increment shortage

The manufacturer decided to call this a 12 control pin because it is the length of the ten control pin (which matches the thickness of the master ring) "plus two" increments.

If it had been four increments difference, a seven cut in this example, the control pin would have needed to be longer than a 10 by four increments, and would have been a 14 control pin.

But what if the opposite situation had occurred? What if the key that was intended to operate at the master ring shearline had a shallower cut than the other key?

In this example, the key intended to operate at the standard shearline had a three cut, so let's examine what would happen if the key intended to operate at the master ring shearline had a number one cut instead.

Because the number one cut is shallower than the number three, it will bring the number three bottom pin up too high, effectively blocking the standard shearline, which is a good thing. But it will ALSO bring the 10 control pin too high for the control sleeve shearline by two increments, blocking that as well, which is not a good thing.
(see fig. 2-4)

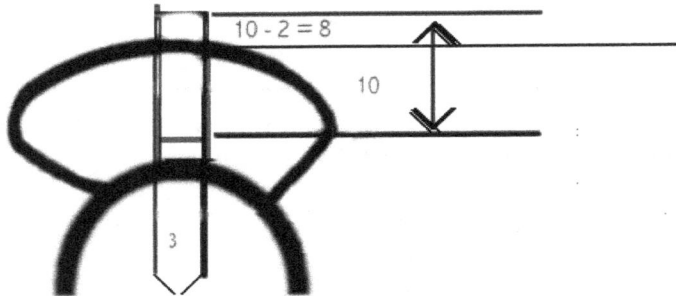

Fig. 2-4) The 8 Control Pin is 2 increments shorter than the 10

So in this case, we need a control pin that is two increments shorter than the 10 control pin. In other words, we need one that is the length of the ten control pin MINUS two increments. The manufacturer, for obvious reasons, calls this an 8 control pin.

If it had been, for example, four increments difference, it would have been a 6 control pin because (10 - 4 = 6). Will it always be an even number? No, it could be a 13 or a 9, for example.

The official procedure for calculating the control pins uses exactly this principle, but in an organized formula.

Because the calculations for the control pin are based on the relationship to the 10 increment thick control sleeve, the calculation begins by creating a CONTROL REFERENCE NUMBER by adding the CONTROL ADDITIVE of 10 to each of the cuts of the control key.

Control Key cuts:	3	5	7	1	2	3
Control Additive:	10	10	10	10	10	10
Control Reference Number:	13	15	17	11	12	13

Write the Control Reference Number (CRN) neatly and leave a few spaces between the numbers.

13   15   17   11   12   13

Below that write the key number intended to operate at the lower (standard) shearline, and write it in the same manner with each cut lining up under the corresponding cut on the other key. Draw a line below that to separate the numbers you just wrote from the results you will get.

16

Small Format (and Mixed Format) Interchangeable Core Servicing

For this example, we will assume no master key is involved, and the individual key is 175875. Therefore, the total of Bottom Pin plus Master Pin will equal the key cuts (or more precisely, the Bottom Pins).

Control Reference Number:	13	15	17	11	12	13
BP + MP	1	7	5	8	7	5

Now calculate the DIFFERENCE between the two in each position.

	13	15	17	11	12	13
	1	7	5	8	7	5
Control Pin:	12	8	12	3	5	8

Because the core fits into a standard cylinder housing, the core itself is smaller than a cylinder. Yet it uses standard size pins and springs. That means that the driver pin must be carefully calculated so as to not crush the spring.

For this reason SFIC products use another pin calculation called a STACK HEIGHT.

In simple terms, a stack height precisely (or nearly precisely) compensates for the length of the pins in a chamber by using a short driver with long pins and a long driver with short pins.

To calculate these for the A2 small format products, we write the total stack height, which for the A2 is 23 for each position, separated by spaces as above. Below it we write the Control Reference Number, and subtract the CRN from the Stack Height to determine the Driver size for each position:

	23	23	23	23	23	23
	13	15	17	11	12	13
Driver:	10	8	6	12	11	10

The nice thing about this calculation is that it only needs to be done once in any customer system, because as long as the CONTROL KEY (the key intended to remove the core from the housing by operating at the Control Sleeve shearline) does not change, neither will the drivers for the system.

In the example presented here, we assumed there was no master key. Master keying the interchangeable core requires the use of the "split pins" because the upper shearline is reserved for the control key to pull or insert the core, and does not turn far enough to act as an operating key.

Now let's do the same core calculations again, but also with a master key of 353431 involved:

Step One: Write the two combinations that will operate at the standard operating (lower) shearline. Write neatly and leave a few spaces between positions to avoid errors:

3	5	3	4	3	1
1	7	5	8	7	5

Step Two: Write the smallest number of each pair below them. This will be the Bottom Pin.

	3	5	3	4	3	1
	1	7	5	8	7	5
BP	1	5	3	4	3	1

Step Three: Write the difference between the two below that. This will be the Master Pin (MP). Bear in mind that you are not exactly subtracting. From a mathematical standpoint you are finding the "absolute difference" [ABS( ) function] between the two. If the bottom number is larger than the upper number, you would subtract the upper number from the bottom, rather than dealing with normal subtraction procedures. This will be the Master pin (MP) If there is no difference, standard procedures are to either leave it blank or put a symbol such as an asterisk (*) as a place holder. Do NOT write a zero. There is no zero master pin!

	3	5	3	4	3	1
	1	7	5	8	7	5
BP	1	5	3	4	3	1
MP	2	2	2	4	4	4

Step Four: Determine the total of pins for each position, either by adding the bottom and master pins or by simply writing the larger of each pair for that position (shown in parenthesis in the example below).

	(3)	5	3	4	3	1
	1	(7)	(5)	(8)	(7)	(5)
BP	1	5	3	4	3	1
MP	2	2	2	4	4	4
(total)	3	7	5	8	7	5

This will be the number you use at the lower position in calculating the control pins.

Step Five: Calculate the Control Reference Number for the customer's system. (In this case, we actually already did it above. As long as our control key cuts do not change we do not need to re-calculate this, but we will copy it to here to keep the steps in order.)

The CONTROL REFERENCE NUMBER is created by adding the CONTROL ADDITIVE of 10 to each of the cuts of the control key.

Control Key cuts	3	5	7	1	2	3
Control Additive	10	10	10	10	10	10
CRN	13	15	17	11	12	13

Step Six: Calculate the Control Pins.

Write the Control Reference Number (CRN) neatly and leave a few spaces between the numbers.

          13   15   17   11   12   13

Below that write the key number intended to operate at the lower (standard) shearline, (deepest cut for each position or total of bottom pin plus master pin) and write it in the same manner with each cut lining up under the corresponding cut on the other key. Draw a line below that to separate the numbers you just wrote from the results you will get. For this example, we calculated this above to be 375875.

Control Reference Number	13	15	17	11	12	13
deepest or total BP + MP	3	7	5	8	7	5

Now calculate the DIFFERENCE between the two in each position.

Control Reference Number:	13	15	17	11	12	13
deepest or total BP + MP:	3	7	5	8	7	5
Control Pin:	10	8	12	3	5	8

Step Seven: Calculate the driver lengths for the customer's system. Once again, we already did this in the previous example, and because the control key has not changed, would not have to do it here, but we are doing so just to keep the steps in order. These are the drivers for every core in this customer's system as long as the control key remains the same. To calculate these for the A2 small format products, we write the total stack height, which for the A2 is 23 for each position, separated by spaces as above. Below it we write the Control Reference Number, and subtract the CRN from the Stack Height to determine the Driver size for each position:

	23	23	23	23	23	23
	13	15	17	11	12	13
Driver:	10	8	6	12	11	10

That is it! Just seven simple steps to fully calculate the pinning for a customer's system, and two of them only need done once, so it is really just five simple steps:

1) Write the two combinations that will operate at the standard operating (lower) shearline.

2) Write the smallest number of each pair below them (BP).

3) Write the difference between the two below that (MP).

4) Determine the total of pins for each position

5) Calculate the Control Reference Number for the customer's system.
   (Control Additive plus Control Key Cuts)
   ** This is once per customer control key **

6) Calculate the Control Pins.

7) Calculate the driver lengths for the customer's system.
   ** This is once per customer control key **

A2 Dimensions:

A2 Key Bitting Depths		Bottom Pin Lengths		Top Pin Lengths	
0	.318"	0A	.110"	2B	.025"
1	.3055"	1A	.1225"	3B	.0375"
2	.293"	2A	.135"	4B	.050"
3	.2805"	3A	.1475"	5B	.0625"
4	.268"	4A	.160"	6B	.075"
5	.2555"	5A	.1725"	7B	.0875"
6	.243"	6A	.185"	8B	.100"
7	.2305"	7A	.1975"	9B	.1125"
8	.218"	8A	.210"	10B	.125"
9	.2055"	9A	.2225"	11B	.1375"
				12B	.150"
		19A = .3475" special		13B	.1625"
		for 9 cut to		14B	.175"
		upper		15B	.1875"
		shearline only		16B	.200"
				17B	.2125"
				18B	.225"
				19B	.2375"

Note: Above figures are usually rounded off on most charts.

Control Additive for A2 = 10
Stack Height for A2 = 23 for standard Products using standard SFIC springs.
Control Reference Number = (Control Additive plus Control Key Cut)
Control Reference number – (total of bottom pin plus master pin) = **Control Pin**
Stack Height minus CRN = Driver Pin (Always the same for a given control key)

The following is a Universal A2 pinning calculator. You manually select the bottom pin and master pin in the usual manner. Then to use it for any given position on the core, calculate the total of Bottom Pin plus Master Pin (or the deepest of the two) for that position. Find that value in the left column. Then trace that row to the right until you are under the column matching the Control Key Cut for that position. The number where the row and column meet is the Control Pin for that position. Now trace down that column to the Driver pin calculation. Repeat for each additional position.

Control Key Cut

Total of Bottom plus Master Pin	0	1	2	3	4	5	6	7	8	9
0	10	11	12	13	14	15	16	17	18	19
1	9	10	11	12	13	14	15	16	17	18
2	8	9	10	11	12	13	14	15	16	17
3	7	8	9	10	11	12	13	14	15	16
4	6	7	8	9	10	11	12	13	14	15
5	5	6	7	8	9	10	11	12	13	14
6	4	5	6	7	8	9	10	11	12	13
7	3	4	5	6	7	8	9	10	11	12
8	2	3	4	5	6	7	8	9	10	11
9		2	3	4	5	6	7	8	9	10
Driver	13	12	11	10	9	8	7	6	5	4

Notes:

Small Format (and Mixed Format) Interchangeable Core Servicing

Chapter Three:
The A3 Small Format Interchangeable Core

These locks use a removable core which contains the bottom pins, master pins (if any), the drivers and springs and an additional pin between the master pins and drivers that can be referred to as a build-up pin or a control pin. They operate on a principle similar to the Corbin-Russwin master ring cylinder, except that on these the ring only goes part of the way around the plug, and serves the function of holding the core in the housing. Also, because the upper shearline is used to operate the control lug which holds the core in place or releases it, it only can be pinned one way, making it simpler.

The beauty of the principle behind this lock cylinder is that lining some of the pins up at the standard shearline and some of the pins up at the upper control shearline accomplishes NOTHING! All of the pins must be lined up at one shearline or the other in order to turn the key. Because of the design of the control sleeve, it can only rotate a quarter turn and cannot actually operate the lock.

The upper shearline is used to move the portion of the ring that is used as a control lug to hold the core in the cylinder, thus releasing the core from its housing when turned one eighth of a turn. Figure 3-1 shows the parts that make up a typical small format interchangeable core.

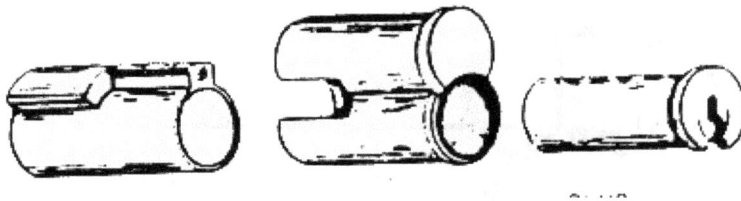

Fig 3-1) Parts of an SFIC core

The control sleeve is rotated when a set of pins blocks the lower (standard) shearline, and lines up at the upper (Control Sleeve) shearline. These pins are generally called a BUILD-UP PIN in most other types of locks, but are referred to as a CONTROL PIN in the Small Format (and mixed format) Interchangeable Core products. It makes up for the difference between the Bottom Pin (and the Master Pin if it exists) and the Control Sleeve shearline when the key is inserted that should operate at the upper (control sleeve) shearline.

To understand this better, we will use an example of just one chamber and no extra master pins (split pin method) needed.

If we have a three cut on both the master and individual key in a specific position, the individual key will bring the number three Bottom Pin to the operating shearline.

But if we had another key with the same cut in that position that we wanted to operate at the upper shearline as well, it would bring the Bottom Pin to the standard shearline, but we would still need to make up the distance from there to the control sleeve shear line.

That means the build-up pin (control pin) has to be the exact thickness of the control sleeve because there is no difference between the two keys. The thickness of the control sleeve is approximately .125 inches (one eighth of an inch). Therefore, in order for our second key to line up pins at the upper shearline as well, it would need a control pin with a length of .125".

The key system referred to as the A3 system uses depths of 0 to 6 with increments of approximately .0175 (seventeen and a half thousandths of an inch each). The .125 is approximately 7 of the A3 increments in thickness.

(Despite published and test data the actual control sleeve thickness is usually closer to .123")

Fig 3-2) The 7 control pin is the thickness of the control sleeve

Now we need to know what to call such a pin. Well, because it is the length of control pin that will be used whenever there is no difference between the two keys, and is equal to the thickness of the control sleeve which is equal to 7 increments, why not call it a 7 control pin? That is exactly what the manufacturer decided as well.

But what if they were not the same? This is quite likely, because if every chamber had a 7 control pin needed, there would be no way to make it operate one shearline or the other. Both shearlines would be free spinning when the key was inserted, which would be an undesirable outcome.

So it is likely that few of the chambers would actually use the same cuts on both the operating key and the control key.

24

What if the cut on the second key had been two increments deeper, in this example a five cut?

Now instead of bringing the three bottom pin to the standard shearline, it brings it two increments too short for the standard shearline, which allows the control pin to drop into the plug by two increments, effectively blocking that shearline.

That is actually a good thing, because we WANT our two keys to operate at different shearlines, not open both. But it also means that the 7 control pin will not reach the control sleeve shear line either. It is too short by two increments, so instead of a 7 we need one that is two increments longer.

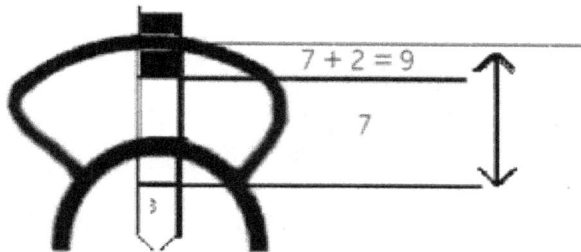

Fig 3-3)   The 9 Control Pin makes up for the 2 increment shortage

The manufacturer decided to call this a 9 control pin because it is the length of the seven control pin (which matches the thickness of the master ring) "plus two" increments.

If it had been four increments difference, a seven cut in this example, the control pin would have needed to be longer than a 7 by four increments, and would have been an 11 control pin.

But what if the opposite situation had occurred? What if the key that was intended to operate at the master ring shearline had a shallower cut than the other key?

In this example, the key intended to operate at the standard shearline had a three cut, so let's examine what would happen if the key intended to operate at the master ring shearline had a number one cut instead.

Because the number one cut is shallower than the number three, it will bring the number three bottom pin up too high, effectively blocking the standard shearline, which is a good thing. But it will ALSO bring the 7 control pin too high for the control sleeve shearline by two increments, blocking that as well, which is not a good thing.
(see fig. 3-4)

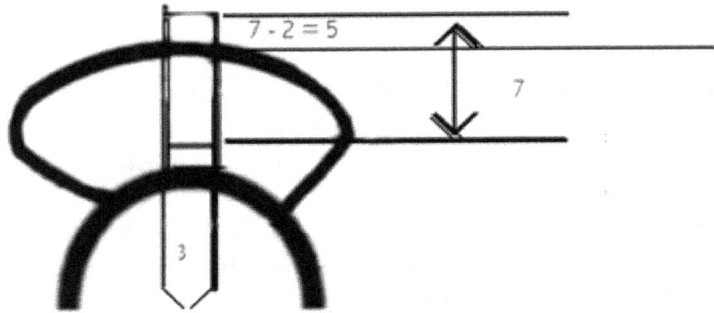

Fig. 3-4) The 5 Control Pin is 2 increments shorter than the 7

So in this case, we need a control pin that is two increments shorter than the 7 control pin. In other words, we need one that is the length of the seven control pin MINUS two increments. The manufacturer, for obvious reasons, calls this an 5 control pin.

If it had been, for example, four increments difference, it would have been a 3 control pin because (7 - 4 = 3).

The official procedure for calculating the control pins uses exactly this principle, but in an organized formula.

Because the calculations for the control pin are based on the relationship to the 7 increment thick control sleeve, the calculation begins by creating a CONTROL REFERENCE NUMBER by adding the CONTROL ADDITIVE of 7 to each of the cuts of the control key.

	3	5	6	1	2	3
Control Additive	7	7	7	7	7	7
CRN	10	12	13	8	9	10

Write the Control Reference Number (CRN) neatly and leave a few spaces between the numbers.

CRN	10	12	13	8	9	10

Below that write the key number intended to operate at the lower (standard) shearline, and write it in the same manner with each cut lining up under the corresponding cut on the other key. Draw a line below that to separate the numbers you just wrote from the results you will get.

26

For this example, we will assume no master key is involved, and the individual key is 165345. Therefore, the total of Bottom Pin plus Master Pin will equal the key cuts (or more precisely, the Bottom Pins).

CRN	10	12	13	8	9	10
	1	6	5	3	4	5

Now calculate the DIFFERENCE between the two in each position.

CRN	10	12	13	8	9	10
	1	6	5	3	4	5
Control Pin:	9	6	8	5	5	5

Because the core fits into a standard cylinder housing, the core itself is smaller than a cylinder. Yet it uses standard size pins and springs. That means that the driver pin must be carefully calculated so as to not crush the spring.

For this reason SFIC products use another pin calculation called a STACK HEIGHT.

In simple terms, a stack height precisely (or nearly precisely) compensates for the length of the pins in a chamber by using a short driver with long pins and a long driver with short pins.

To calculate these for the A3 series products, we write the total stack height, which is 16 for each position, separated by spaces as above. Below it we write the Control Reference Number, and subtract the CRN from the Stack Height to determine the Driver size for each position:

Stack Height:	16	16	16	16	16	16
Control Reference Number	10	12	13	8	9	10
Driver:	6	4	3	8	7	6

The nice thing about this calculation is that it only needs to be done once in any customer system, because as long as the CONTROL KEY (the key intended to remove the core from the housing by operating at the Control Sleeve shearline) does not change, neither will the drivers for the system.

In the example presented here, we assumed there was no master key. Master keying the interchangeable core requires the use of the "split pins" because the upper shearline is reserved for the control key to pull or insert the core, and does not turn far enough to act as an operating key.

Now let's do the same core calculations again, but also with a master key of 353431 involved:

Step One: Write the two combinations that will operate at the standard operating (lower) shearline. Write neatly and leave a few spaces between positions to avoid errors:

3	5	3	4	3	1
1	6	5	3	4	5

Step Two: Write the smallest number of each pair below them. This will be the Bottom Pin (BP).

	3	5	3	4	3	1
	1	6	5	3	4	5
BP	1	5	3	3	3	1

Step Three: Write the difference between the two below that. Bear in mind that you are not exactly subtracting. From a mathematical standpoint you are finding the "absolute difference" [ABS( ) function] between the two. If the bottom number is larger than the upper number, you would subtract the upper number from the bottom, rather than dealing with normal subtraction procedures. This will be the Master pin (MP) If there is no difference, standard procedures are to either leave it blank or put a symbol such as an asterisk (*) as a place holder. Do NOT write a zero. There is no zero master pin!

	3	5	3	4	3	1
	1	6	5	3	4	5
BP	1	5	3	3	3	1
MP	2	1	2	1	1	4

Step Four: Determine the total of pins for each position, either by adding the bottom and master pins or by simply writing the larger of each pair for that position (in parenthesis in the example below).

	(3)	5	3	(4)	3	1
	1	(6)	(5)	3	(4)	(5)
BP:	1	5	3	3	3	1
MP:	2	1	2	1	1	4
total BP + MP:	3	6	5	4	4	5

This will be the number you use at the lower position in calculating the control pins.

Step Five: Calculate the Control Reference Number for the customer's system. (In this case, we actually already did it above. As long as out control key cuts do not change we do not need to re-calculate this, but we will copy it to here to keep the steps in order.)

28

The CONTROL REFERENCE NUMBER is created   by adding the CONTROL ADDITIVE of 7 to each of the cuts of the control key.

Control Key cuts	3	5	6	1	2	3
Control Additive	7	7	7	7	7	7
CRN	10	12	13	8	9	10

Step Six: Calculate the Control Pins.

Write the Control Reference Number (CRN) neatly and leave a few spaces between the numbers.

CRN	10	12	13	8	9	10

Below that write the key number intended to operate at the lower (standard) shearline, (deepest cut for each position or total of bottom pin plus master pin) and write it in the same manner with each cut lining up under the corresponding cut on the other key. Draw a line below that to separate the numbers you just wrote from the results you will get. For this example, we calculated this above to be 375875.

CRN	10	12	13	8	9	10
deepest:	3	6	5	4	4	5

Now calculate the DIFFERENCE between the two in each position.

CRN	10	12	13	8	9	10
deepest:	3	6	5	4	4	5
Control Pin:	8	6	8	4	5	5

Step Seven: Calculate the driver lengths for the customer's system. Once again, we already did this in the previous example, and because the control key has not changed, would not have to do it here, but we are doing so just to keep the steps in order. To calculate these for the A3, we write the total stack height, which for A3 is 16 for each position, separated by spaces as above. Below it we write the Control Reference Number, and subtract the CRN from the Stack Height to determine the Driver size for each position:

Stack Height:	16	16	16	16	16	16
CRN:	10	12	13	8	9	10
Driver:	6	4	3	8	7	6

These are the drivers for every core in this customer's system as long as the control key remains the same.

That is it! Just seven simple steps to fully calculate the pinning for a customer's system, and two of them only need done once, so it is really just five simple steps:

1) Write the two combinations that will operate at the standard operating (lower) shearline.

2) Write the smallest number of each pair below them (BP).

3) Write the difference between the two below that (MP).

4) Determine the total of pins for each position

5) Calculate the Control Reference Number for the customer's system.
   (Control Additive plus Control Key Cuts)
   ** This is once per customer control key **

6) Calculate the Control Pins.

7) Calculate the driver lengths for the customer's system.
   ** This is once per customer control key **

A3 Dimensions:

A3 Key Bitting Depths		Bottom Pin Lengths		Top Pin Lengths	
0	.318	0C	.110"	2D	.036"
1	.300	1C	.128"	3D	.054"
2	.282	2C	.146"	4D	.072"
3	.264	3C	.164"	5D	.090"
4	.246	4C	.182"	6D	.108"
5	.228	5C	.200"	7D	.126"
6	.210	6C	.218"	8D	.144
				9D	.162"
				10D	.180"
				11D	.198"
				12D	.216"
				13D	.234"

Note: Above figures usually rounded off on most charts.

Control Additive for A3 = 7
Stack Height for A3 = 16
Control Reference number= (Control Additive plus Control Key Cut)
Control Reference number – (total of bottom pin plus master pin) = **Control Pin**
Stack Height minus CRN = **Driver Pin** (Always the same for a given control key)

The following is a Universal A3 pinning calculator. You manually select the bottom pin and master pin in the usual manner. Then to use it for any given position on the core, calculate the total of Bottom Pin plus Master Pin (or the deepest of the two) for that position. Find that value in the left column. Then trace that row to the right until you are under the column matching the Control Key Cut for that position. The number where the row and column meet is the Control Pin for that position. Now trace down that column to the Driver pin calculation. Repeat for each additional position.

SFIC A3

Bottom Pin + Master Pin	Control Key						
	0	1	2	3	4	5	6
0	7	8	9	10	11	12	13
1	6	7	8	9	10	11	12
2	5	6	7	8	9	10	11
3	4	5	6	7	8	9	10
4	3	4	5	6	7	8	9
5	2	3	4	5	6	7	8
6	1	2	3	4	5	6	7
Driver Pins							
	9	8	7	6	5	4	3

31

Notes:

Chapter Four:
The A4 Small Format Interchangeable Core

These locks use a removable core which contains the bottom pins, master pins (if any), the drivers and springs and an additional pin between the master pins and drivers that can be referred to as a build-up pin or a control pin. They operate on a principle similar to the Corbin-Russwin master ring cylinder, except that on these the ring only goes part of the way around the plug, and serves the function of holding the core in the housing. Also, because the upper shearline is used to operate the control lug which holds the core in place or releases it, it only can be pinned one way, making it simpler.

The beauty of the principle behind this lock cylinder is that lining some of the pins up at the standard shearline and some of the pins up at the upper control shearline accomplishes NOTHING! All of the pins must be lined up at one shearline or the other in order to turn the key. Because of the design of the control sleeve, it can only rotate a quarter turn and cannot actually operate the lock.

The upper shearline is used to move the portion of the ring that is used as a control lug to hold the core in the cylinder, thus releasing the core from its housing when turned one eighth of a turn. Figure 4-1 shows the parts that make up a typical small format interchangeable core.

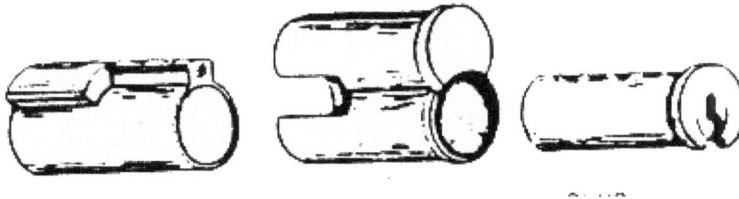

Fig 4-1) Parts of a Best (SFIC) core

The control sleeve is rotated when a set of pins blocks the lower (standard) shearline, and lines up at the upper (Control Sleeve) shearline. These pins are generally called a BUILD-UP PIN in most other types of locks, but are referred to as a CONTROL PIN in the Small Format (and mixed format) Interchangeable Core products. It makes up for the difference between the Bottom Pin (and the Master Pin if it exists) and the Control Sleeve shearline when the key is inserted that should operate at the upper (control sleeve) shearline.

To understand this better, we will use an example of just one chamber and no extra master pins (split pin method) needed.

If we have a three cut on both the master and individual key in a specific position, the individual key will bring the number three Bottom Pin to the operating shearline.

But if we had another key with the same cut in that position that we wanted to operate at the upper shearline as well, it would bring the Bottom Pin to the standard shearline, but we would still need to make up the distance from there to the control sleeve shear line.

That means the build-up pin (control pin) has to be the exact thickness of the control sleeve because there is no difference between the two keys. The thickness of the control sleeve is approximately .125 inches (one eighth of an inch). Therefore, in order for our second key to line up pins at the upper shearline as well, it would need a control pin with a length of .125".

The key system referred to as the A4 system uses depths of 0 to 5 with increments of approximately .0205 (twenty and a half thousandths of an inch each). The .125 is approximately 6 of the A4 increments in thickness.

(Despite published and test data the actual control sleeve thickness is closer to .123")

Fig 4-2) The 6 Control Pin is the thickness of the control sleeve.

Now we need to know what to call such a pin. Well, because it is the length of control pin that will be used whenever there is no difference between the two keys, and is equal to the thickness of the control sleeve which is equal to 6 increments, why not call it a 6 control pin? That is exactly what the manufacturer decided as well.

But what if they were not the same? This is quite likely, because if every chamber had a 6 control pin needed, there would be no way to make it operate one shearline or the other. Both shearlines would be free spinning when the key was inserted, which would be an undesirable outcome.

So it is likely that few of the chambers would actually use the same cuts on both the operating key and the control key.

What if the cut on the second key had been two increments deeper, in this example a five cut?

Now instead of bringing the three bottom pin to the standard shearline, it brings it two increments too short for the standard shearline, which allows the control pin to drop into the plug by two increments, effectively blocking that shearline.

That is actually a good thing, because we WANT our two keys to operate at different shearlines, not open both. But it also means that the 6 control pin will not reach the control sleeve shear line either. It is too short by two increments, so instead of a 6 we need one that is two increments longer.

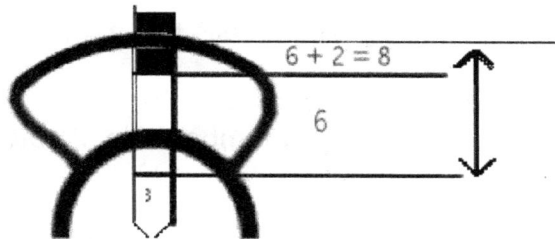

Fig 4-3) The 8 pin is 2 increments longer than the 6.

The manufacturer decided to call this an 8 control pin because it is the length of the six control pin (which matches the thickness of the master ring) "plus two" increments.

If it had been four increments difference, a seven cut in this example, the control pin would have needed to be longer than a 6 by four increments, and would have been a 10 control pin.

But what if the opposite situation had occurred? What if the key that was intended to operate at the master ring shearline had a shallower cut than the other key?

In this example, the key intended to operate at the standard shearline had a three cut, so let's examine what would happen if the key intended to operate at the master ring shearline had a number one cut instead.

Because the number one cut is shallower than the number three, it will bring the number three bottom pin up too high, effectively blocking the standard shearline, which is a good thing. But it will ALSO bring the 6 control pin too high for the control sleeve shearline by two increments, blocking that as well, which is not a good thing.

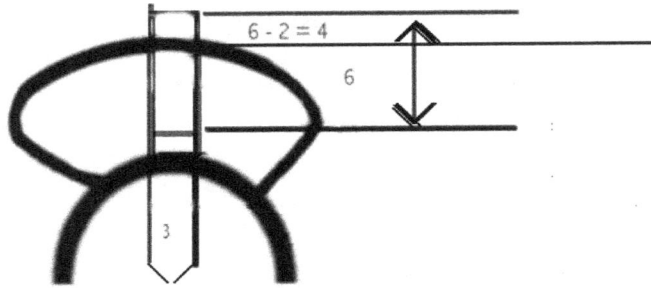

Fig. 4-4) The four pin is 2 increments shorter than the 6

So in this case, we need a control pin that is two increments shorter than the 6 control pin. In other words, we need one that is the length of the six control pin MINUS two increments. The manufacturer, for obvious reasons, calls this a 4 control pin.

If it had been, for example, four increments difference, it would have been a 2 control pin because (6 - 4 = 2). Will it always be an even number? No, it could be an 11 or a 3, for example.

The official procedure for calculating the control pins uses exactly this principle, but in an organized formula.

Because the calculations for the control pin are based on the relationship to the 6 increment thick control sleeve, the calculation begins by creating a CONTROL REFERENCE NUMBER by adding the CONTROL ADDITIVE of 6 to each of the cuts of the control key.

Control Key cuts	3	5	4	1	2	3
Control Additive	6	6	6	6	6	6
CRN	9	11	10	7	8	9

Write the Control Reference Number (CRN) neatly and leave a few spaces between the numbers.

CRN	9	11	10	7	8	9

Below that write the key number intended to operate at the lower (standard) shearline, and write it in the same manner with each cut lining up under the corresponding cut on the other key. Draw a line below that to separate the numbers you just wrote from the results you will get. For this example, we will assume no master key is involved, and the individual key is 125345. Therefore, the total of Bottom Pin plus Master Pin will equal the key cuts (or more precisely, the Bottom Pins).

Control Reference Number:	9	11	10	7	8	9
BP + MP:	1	2	5	3	4	5

Now calculate the DIFFERENCE between the two in each position.

Control Reference Number:	9	11	10	7	8	9
BP + MP:	1	2	5	3	4	5
Control Pin:	8	9	5	4	4	4

Because the core fits into a standard cylinder housing, the core itself is smaller than a cylinder. Yet it uses standard size pins and springs. That means that the driver pin must be carefully calculated so as to not crush the spring.

For this reason SFIC products use another pin calculation called a STACK HEIGHT.
In simple terms, a stack height precisely (or nearly precisely) compensates for the length of the pins in a chamber by using a short driver with long pins and a long driver with short pins.

To calculate these for the A4 series products, we write the total stack height, which is 14 for each position, separated by spaces as above. Below it we write the Control Reference Number, and subtract the CRN from the Stack Height to determine the Driver size for each position:

Control Additive:	14	14	14	14	14	14
Control Reference Number:	9	11	10	7	8	9
Driver:	5	3	4	7	6	5

The nice thing about this calculation is that it only needs to be done once in any customer system, because as long as the CONTROL KEY (the key intended to remove the core from the housing by operating at the Control Sleeve shearline) does not change, neither will the drivers for the system.

In the example presented here, we assumed there was no master key. Master keying the interchangeable core requires the use of the "split pins" because the upper shearline is reserved for the control key to pull or insert the core, and does not turn far enough to act as an operating key.

Now let's do the same core calculations again, but also with a master key of 353431 involved:

Step One: Write the two combinations that will operate at the standard operating (lower) shearline. Write neatly and leave a few spaces between positions to avoid errors:

3	5	3	4	3	1
1	2	5	3	4	5

Step Two: Write the smallest number of each pair below them. This will be the Bottom Pin (BP).

	3	5	3	4	3	1
	1	2	5	3	4	5
BP	1	2	3	3	3	1

Step Three: Write the difference between the two below that. Bear in mind that you are not exactly subtracting. From a mathematical standpoint you are finding the "absolute difference" [ABS( ) function] between the two. If the bottom number is larger than the upper number, you would subtract the upper number from the bottom, rather than dealing with normal subtraction procedures. This will be the Master pin (MP) If there is no difference, standard procedures are to either leave it blank or put a symbol such as an asterisk (*) as a place holder. Do NOT write a zero. There is no zero master pin!

	3	5	3	4	3	1
	1	2	5	3	4	5
BP	1	2	3	3	3	1
MP	2	3	2	1	1	4

Step Four: Determine the total of pins for each position, either by adding the bottom and master pins or by simply writing the larger of each pair for that position.

	(3)	(5)	3	(4)	3	1
	1	2	(5)	3	(4)	(5)
BP	1	2	3	3	3	1
MP	2	3	2	1	1	4
deepest	3	5	5	4	4	5

This will be the number you use at the lower position in calculating the control pins.

Step Five: Calculate the Control Reference Number for the customer's system. (In this case, we actually already did it above. As long as out control key cuts do not change we do not need to re-calculate this, but we will copy it to here to keep the steps in order.)

The CONTROL REFERENCE NUMBER is created by adding the CONTROL ADDITIVE of 6 to each of the cuts of the control key.

Control Key cuts:	3	5	4	1	2	3
Control Additive:	6	6	6	6	6	6
Control Reference Number:	9	11	10	7	8	9

Step Six: Calculate the Control Pins.

Write the Control Reference Number (CRN) neatly and leave a few spaces between the numbers.

CRN   9   11   10   7   8   9

Below that write the key number intended to operate at the lower (standard) shearline, (deepest cut for each position or total of bottom pin plus master pin) and write it in the same manner with each cut lining up under the corresponding cut on the other key. For this example, we calculated this above to be 355445. Draw a line below that to separate the numbers you just wrote from the results you will get.

Control Reference Number:	9	11	10	7	8	9
deepest or (BP + MP):	3	5	5	4	4	5

Now calculate the DIFFERENCE between the two in each position.

Control Reference Number:	9	11	10	7	8	9
deepest or (BP + MP):	3	5	5	4	4	5
Control Pin:	6	6	5	3	4	4

Step Seven: Calculate the driver lengths for the customer's system. Once again, we already did this in the previous example, and because the control key has not changed, would not have to do it here, but we are doing so just to keep the steps in order. To calculate these for the A3, we write the total stack height, which for A3 is 16 for each position, separated by spaces as above. Below it we write the Control Reference Number, and subtract the CRN from the Stack Height to determine the Driver size for each position:

Control Additive:	14	14	14	14	14	14
Control Reference Number:	9	11	10	7	8	9
Driver:	5	3	4	7	6	5

These are the drivers for every core in this customer's system as long as the control key remains the same.

That is it! Just seven simple steps to fully calculate the pining for a customer's system, and two of them only need done once, so it is really just five simple steps:

1) Write the two combinations that will operate at the standard operating (lower) shearline.

2) Write the smallest number of each pair below them (BP).

3) Write the difference between the two below that (MP).

4) Determine the total of pins for each position

5) Calculate the Control Reference Number for the customer's system.
(Control Additive plus Control Key Cuts)
        ** This is once per customer control key **

6) Calculate the Control Pins.

7) Calculate the driver lengths for the customer's system.
** This is once per customer control key **

A4 Key Bitting Depths		Bottom Pin Lengths		Top Pin Lengths	
0	.318	0E	.110"	1F	.021".
1	.297	1E	.131"	2F	042"
2	.276	2E	.152"	3F	.063"
3	.255	3E	.173"	4F	.084"
4	.234	4E	.194"	5F	.105"
5	.213	5E	.215"	6F	.126"
				7F	.147"
				8F	.168
				9F	.189"
				10F	.210"
				11F	.231"

Note: Above figures may vary -rounded off on most charts.

Control Additive for A4 = 6
Stack Height for A4 = 14
Increment size = .0205 inches
(Control Reference number= Control Additive plus Control Key Cut)
Control Reference number – (total of bottom pin plus master pin) = Control Pin
Stack Height minus CRN = Driver Pin (Always the same for a given control key)

The following is a Universal A4 pinning calculator. You manually select the bottom pin and master pin in the usual manner. Then to use it for any given position on the core, calculate the total of Bottom Pin plus Master Pin (or the deepest of the two) for that position. Find that value in the left column. Then trace that row to the right until you are under the column matching the Control Key Cut for that position. The number where the row and column meet is the Control Pin for that position. Now trace down that column to the Driver pin calculation. Repeat for each additional position.

SFIC A4

	Control Key Cut					
	0	1	2	3	4	5
0	6	7	8	9	10	11
1	5	6	7	8	9	10
2	4	5	6	7	8	9
3	3	4	5	6	7	8
4	2	3	4	5	6	7
5	1	2	3	4	5	6
Driver Pins						
	8	7	6	5	4	3

(Left axis label: Bottom Pin + Master Pin)

Notes:

Chapter Nine
Small Format Core Specifics

These are just some examples of the similarities and differences among the Small Format Interchangeable Core. This listing is not complete because new manufacturers of SFIC products come on the market regularly due to its popularity in the marketplace, particularly in the U.S. and Canada.

Arrow:

Product line is ChoICe modular interchangeable core system.
At this time there are several levels:

- ChoICe Base
    - Similar to Best core but uses a saddle instead of a control sleeve, but for the same purpose.
    - Available in all "standard" (Best) keyways A through M and Q, as well as in Arrow's own 1C, 1D and restricted keyways
    - Six pin and 7 pin available .150 spacing
    - Available in A2 and A4 nut furnished A2 by default.
    - Available in drill resistant version
    - Face plate is removable and replaceable to achieve the desired finish without carrying double the number of cores or more.
    - U.S. Patent # 6,079,240

- Arrow ChoICe Flex (formerly Flexcore)

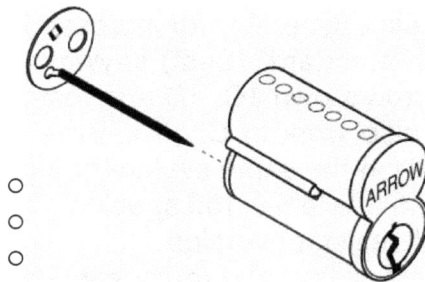

    -
    -
    -
    -
    - A restricting pin extends along the bottom of the keyway into a hole in the blank. Keys with the hole can enter both standard keyway cores and flex pin keyway cores.
    - Arrow keyways 51, 52, 53, 54, 61, 81, 83, 84, 91 only
    - All keyways restricted and issued by contract only based on geographical availability. Without a contract you cannot purchase them.

o Lock can be assembled without restricting pin to create two separate levels of security in the same system, but this is a special factory request only.
o Six pin and 7 pin available .140 spacing
o Available in drill resistant version
o Face plate is removable and replaceable to achieve the desired finish without carrying double the number of cores or more.
o Furnished A2 only.
o U.S. Patents # 5,778,712 and 6,079,240

- Arrow ChoICe Plus

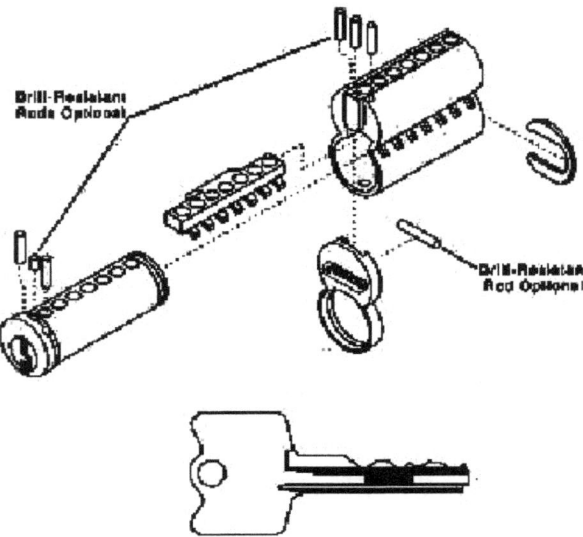

o SImilar to Best core but uses a saddle instead of a control sleeve, but for the same purpose.
o Adds a "finger pin" assembly for increased security
o Available in all "standard" (Best) keyways A through M and Q, as well as in Arrow's own 1C, 1D and restricted keyways
o Plus Key can also work in ChoICe Base cores, permitting two levels of security in the same system or a phase-in approach.
o Six pin and 7 pin available .150 spacing
o Available in drill resistant version
o Face plate is removable and replaceable to achieve the desired finish without carrying double the number of cores or more.
o Furnished A2 only.
o U.S. Patent # 6,079,240. Other patents pending until February 2021.

- Arrow ChoICe HS

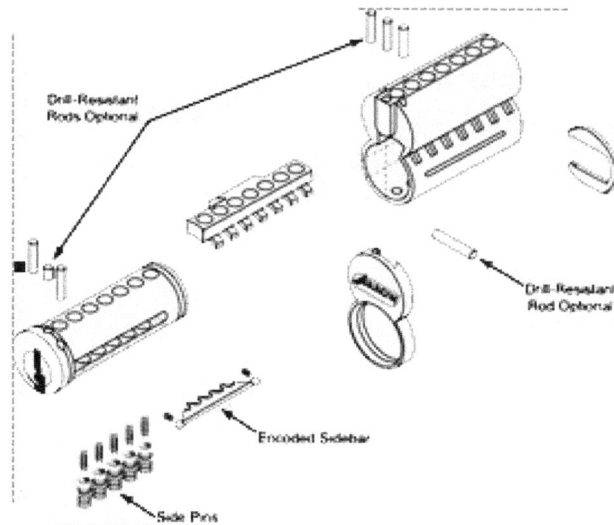

- o Similar to Best core but uses a saddle instead of a control sleeve, but for the same purpose.
- o Uses a five pin sidebar for better security
- o Each side pin position has five possible depths, yielding 3,125 theoretical side bittings per keyway
- o Available in restricted keyways only
- o 7 pin only .140 spacing using A2 depths
- o Available in drill resistant version
- o 40 is standard keyway furnished, not compatible with other SFIC keyways.
- o Face plate is removable and replaceable to achieve the desired finish without carrying double the number of cores or more.
- o U.S. Patent # 6,079,240 and #6,301,942 until July 2018.

- o Arrow ChoICe Pointe
    - o Manufactured to close tolerances by Medeco ™
    - o Available in all "standard" (Best) keyways A through M and Q, as well as in Arrow's own 1C, 1D and restricted keyways
    - o Six pin and 7 pin available .150 spacing
    - o Available in drill resistant version in A2 or A4 with A2 as standard.
    - o Face plate is removable and replaceable to achieve the desired finish without carrying double the number of cores or more.
    - o U.S. Patent # 6,079,240

45

BEST

In 1923, Frank E. Best invented a new type of padlock. He continued to make modifications to it and was granted new patents in 1924 and 1926. Figure 3-1 shows the 1923 patent drawing and the 1924 and 1926 modifications.

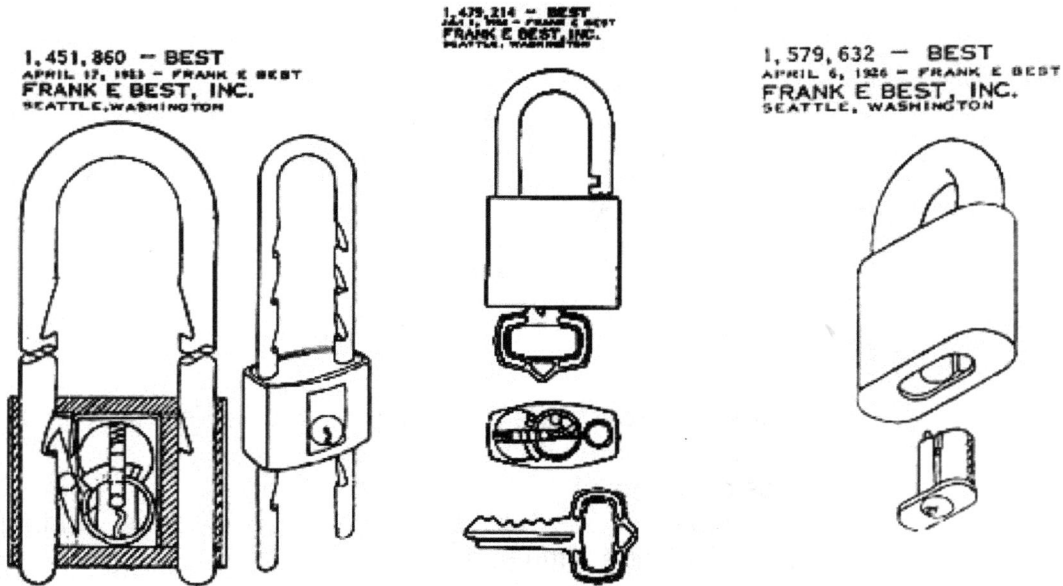

fig 5-Best--1 The original Best lock patent drawings.

Best uses a numbering system for their cores that ends in " –A1" if the core is uncombinated (no pins inserted – no keys attached), so their first
designation for the combinated ones was " –A2" and early catalogs simply show this designation as meaning combinated.

The numbering system for these cores was zero to nine, which on a six pin core yields 4,096 combinations in a master keyed series, or in a seven pin core yields 16,384 combinations in a master keyed series.

These numbers seemed sufficient for most customers and Best was happy with the results until the U.S. Government approached them and wanted a system for all the locks on their bases and padlocks all over the world.

Best developed the A3 numbering system for them in order to get more combinations. They also sold it to Hospitals and similar institutions. It used larger increments, so that it could theoretically use single increment differences in the master key system to get a much larger system. The A3 uses zero to six numbering. They used the existing cores.

However, Best discovered that the A3 numbering system increments were not sufficient to prevent keys from operating locks they were not intended to when using a single step master key progression once the cores and keys began to show wear, so they changed the master key layout specifications for the A3 to provide proper security for their customers.

They also increased the size of the increments and changed the numbering system to a range of zero to five. They called this the A4 numbering system and it was also sold to Universities and other large systems. These also used the existing cores. This was, and remains, highly successful. Although many manufacturers offer equivalent products to the Best A2 series, few use the A3 or A4 numbering systems.

It should be noted that there is no difference physically between an A2, A3 or A4 core until it is pinned. All three have been available in five pin, six pin and seven pin length, although it is exceedingly rare to see an A3 or A4 system that is not 7 pin (seven cuts per key).

Some commonly seen Best keyways are shown here:
(Not actual size)

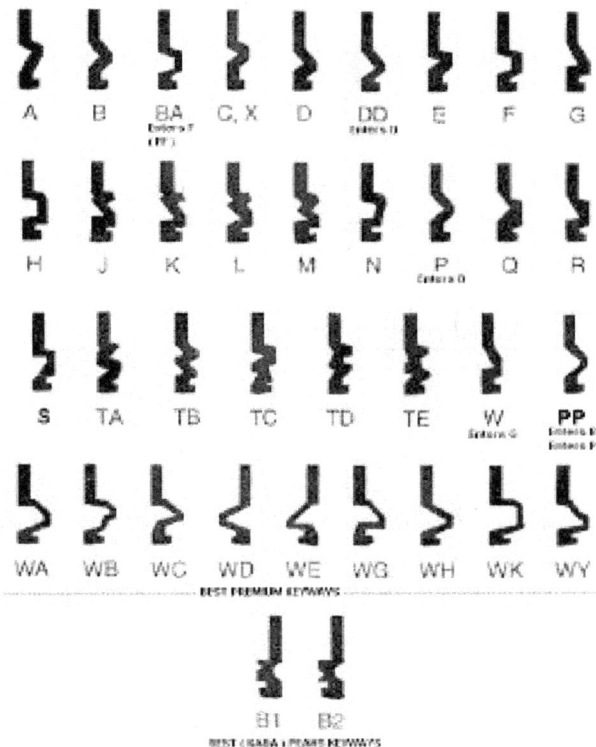

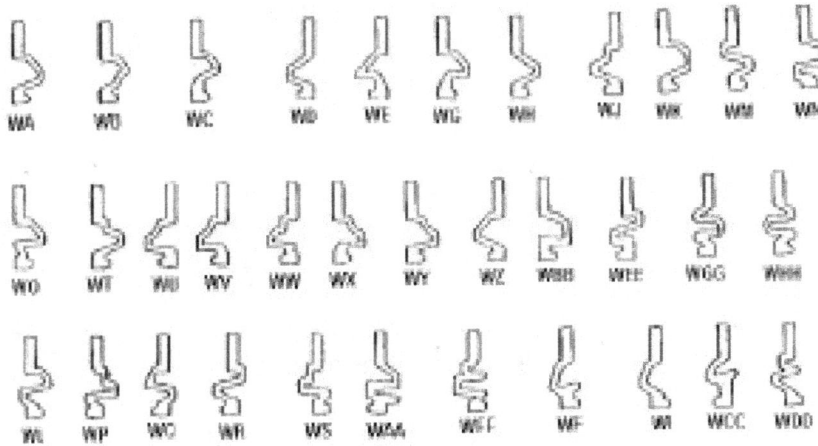

Although BEST has been the model for many manufacturers' SFIC products, it remains very alone in the breadth of its offering for specialty cores.

The 2C core, for example, has an extended plug to permit the core to work easily on the 8L mailbox lock.

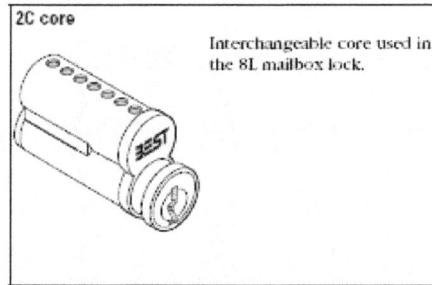

2C core

Interchangeable core used in the 8L mailbox lock.

The 3C core is adapted for the 8E European locksets.

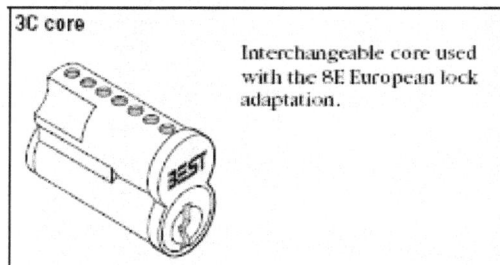

3C core

Interchangeable core used with the 8E European lock adaptation.

The 6C is specially designed for the 4S sliding door lock, which requires a lost motion assembly on the rear of the core.

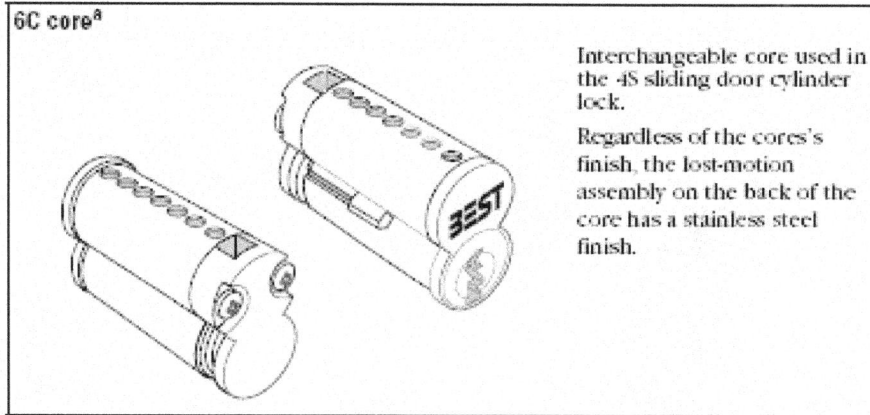

6C core[a]

Interchangeable core used in the 4S sliding door cylinder lock.

Regardless of the cores's finish, the lost-motion assembly on the back of the core has a stainless steel finish.

Best offers the A40305 cores which is prepped for the B26247 spring and the A00127 dust cover.

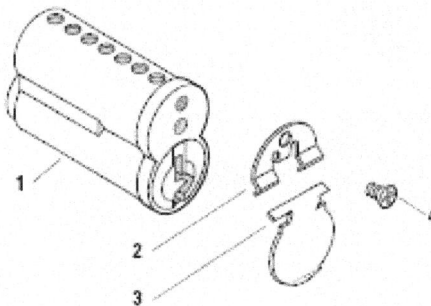

The Best Service Manual is available online and shows each of these in more detail. The above four illustrations are from that manual.

Best offers the 5C core for high security applications, but in order to maintain the UL Rating, it requires factory only pinning. You may not use the UL stamped faceplate if you are doing field pinning.

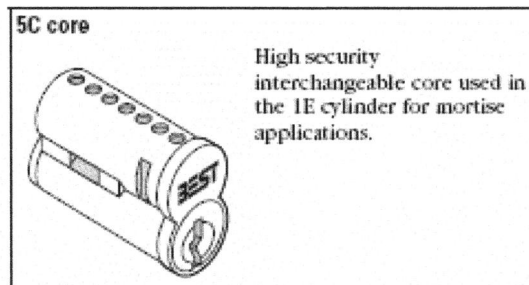

5C core

High security interchangeable core used in the 1E cylinder for mortise applications.

They do make a non-UL stamped faceplate version available for areas where field pinning will be required, but even so, there are some strict requirements. It MUST use hardened stainless steel pins in the chambers closest to the face of the core, and spooled pins in all the other positions except where a 6B pin is required, in which case the standard 6B pin may be used. These requirements also limit the high security to the A2 numbering system.

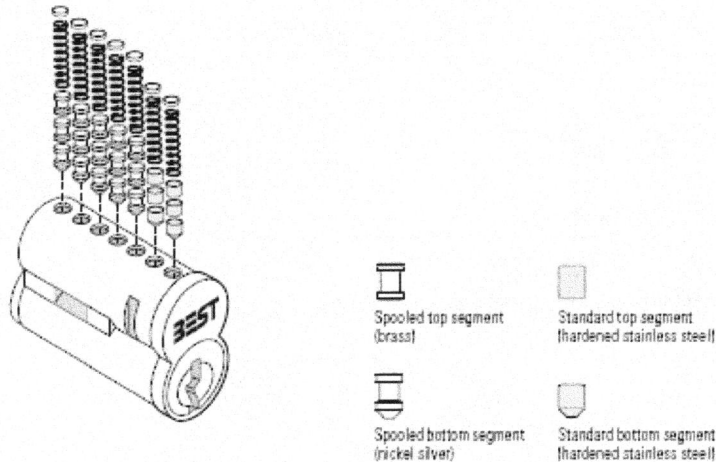

Spooled top segment
(brass)

Standard top segment
(hardened stainless steel)

Spooled bottom segment
(nickel silver)

Standard bottom segment
(hardened stainless steel)

Loading the 5C core for the 1 E7K4 cylinder

From the Best Service Manual for the 5C core

The specialty cores, however, are not interchangeable. Their unique outer shape limits their application.

Sometimes cores that were intended to be temporary during the construction phase are left in place by accident. The chart on the following page may help in finding an appropriate way to replace them.

Construction Keys (There are others as well)

COLOR	KEYWAY	KEY TYPE							
No Paint		Operating –A2	3.	7	8	1	0	7	3
		Control – A2	5	3	0	9	4	5	1
Black	G	Operating –A2	3	7	8	1	0	7	3
		Control – A2	5	3	0	9	4	5	1
Light Blue	F	Operating –A2	3.	7	8	1	0	7	3
		Control – A2	5	3	0	9	4	5	1
Dark Blue	A	Operating –A2	3.	7	8	1	0	7	3
		Control – A2	5	3	0	9	4	5	1
Green	D	Operating –A2	3.	7	8	1	0	7	3
		Control – A2	5	3	0	9	4	5	1
Green	F	Operating –A2	5	2	1	4	3	2	2
		Control – A2	1	6	5	8	9	2	6
Green	G	Operating –A2	1	4	7	6	5	6	
		Control – A2	3	8	9	8	7	2	
PEWTER GREY	J	Operating –A2	1	4	7	6	5	6	
		Control – A2	3	8	9	8	7	2	
ORANGE	A	Operating –A2	5	2	1	4	3	2	2
		Control – A2	1	6	5	8	5	2	6
PINK	L	Operating –A2	1	4	7	6	5	6	
		Control – A2	3	8	9	8	7	2	
RED	A	Operating –A2	1	4	7	6	5	6	
		Control – A2	3	8	9	8	7	2	
RED	F	Operating –A2	3.	7	8	1	0	7	3
		Control – A2	5	3	0	9	4	5	1
TAN	E	Operating –A2	1	4	7	6	5	6	
		Control – A2	3	8	9	8	7	2	
WHITE	F	Operating –A2	1	4	7	6	5	6	
		Control – A2	3	8	9	8	7	2	
WHITE	H	Operating –A2	3.	7	8	1	0	7	3
		Control – A2	5	3	0	9	4	5	1
YELLOW	D	Operating –A2	1	4	7	6	5	6	
		Control – A2	3	8	9	8	7	2	
YELLOW	K	Operating –A2	3.	7	8	1	0	7	3
		Control – A2	5	3	0	9	4	5	1

Best   Product line was original interchangeable core.
   o Offers all "standard" (Best) keyways, obviously
   o available in 5, 6 or 7 pin

Best PKS
   o Offered a thicker and larger "patented" PKS key which was unsuccessful because it failed the patent under legal contest, which then became Premier and no longer listed as Patented.

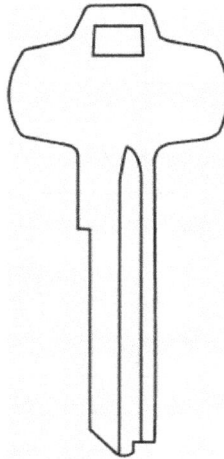

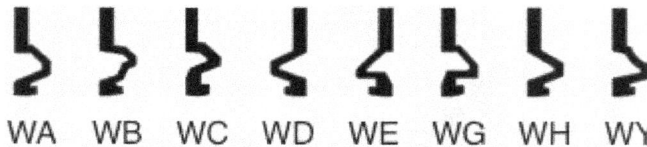

WA   WB   WC   WD   WE   WG   WH   WY

(Not shown actual size)

Best B1-7 Series (KABA Peaks)

   o Best now also offers its own line of Kaba ™ cores and assigned keyways B1 through B7.
   o B3 keyway must be cut on a "grinding wheel cutter" (code machine such as Framon ™ or HPC ™ - cannot be cut with any punch currently on the market.
   o Also has a new higher security core with a check pin and patent protected keyway.

Best MX-8

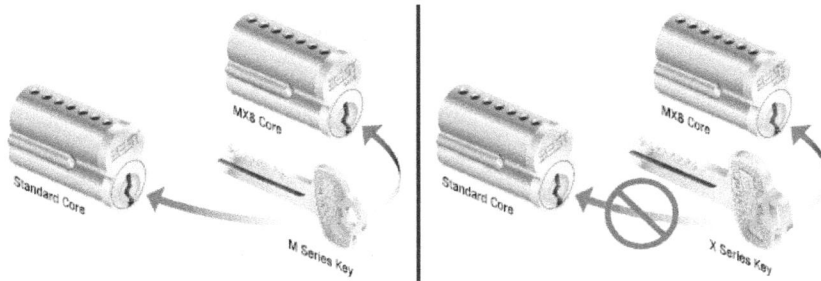

- o has a slider tumbler that must be pushed a specific distance by a cut placed in the side of the key.
- o Two versions of the key are used in the MX8 products. The M Series do not have the slider tumbler cut in them and can only operate cylinders without the slider tumblers. The X Series have the slider tumbler cut and can operate cylinders with and without the slider tumbler.
- o Available in A2, A3 or A4
- o Available in J, K, L & M keyways or one of four special MX-8 only keyways
- o Slider tumbler (similar to check pin) is contained within the plug and has a projection that protrudes into a slot in the shell.

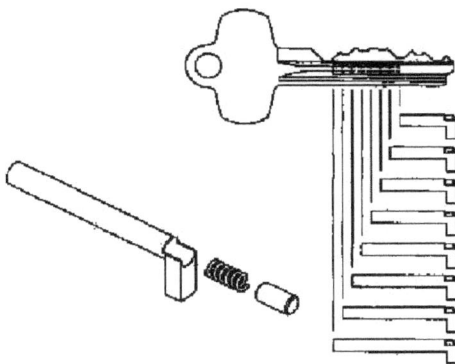

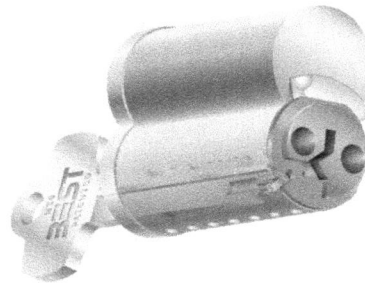

MX8 LOCKED   MX8 UNLOCKED

Falcon

Falcon   Product line has been around since before Best Patent ran out.

- o Tolerances in product line are good
- o Similar to Best core but originally used a brass slide cap instead of individual caps, but for the same purpose.
- o Today is available in both chamber capped and slide capped models.
- o Available in all "standard" keyways as well as some of its own.
- o Sold and stocked by many Locksmith Wholesalers
- o 5, 6 or 7 pin available typically .150 spacing
- o Available in A2 or A4 but A2 is standard.

Falcon Instakey

- o Manufactured two types of core for the Instakey company.
- o Available with standard pinning in Instakey keyway or with segmented pins

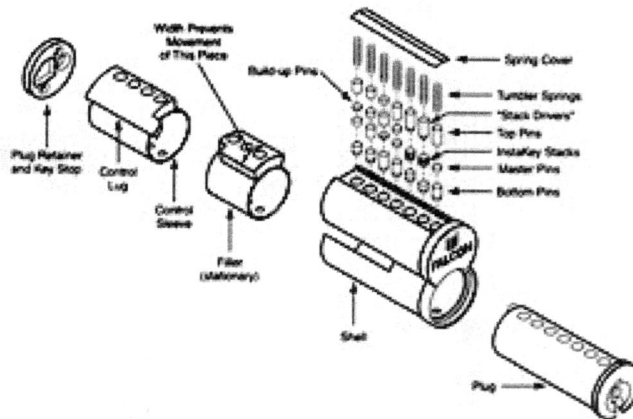

KABA Peaks

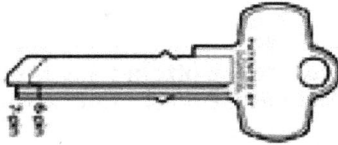

KABA Product line first introduced in the U.S. as Lori Peaks™
- o Uses high quality nickel silver bottom pins to resist wear
- o Uses a "finger pin" activated by "peak" on key for increased security.
- o New model uses improved finger pin with conical bottom to reduce wear problems.
- o Also has higher security model using a sidebar available.
- o Was the first in the U.S. to use a finger pin
- o Several lines available, all including restricted keyways at various levels
- o Has six pin .140 spacing versions that can be pinned using special (J for A2 or K for A4) pins to operate as a cylindrical leverset cylinder, a Schlage ™ knobset, a Corbin-Russwin™ knob cylinder, a Corbin-Russwin™ Large Format removable core or a Sargent ™ Large Format removable core.
- o Large format versions are pinned as though they were small format cores except for using special length bottom pins, and feature a control sleeve that covers all chamber positions (6 or 7)
- o Available in A2 or A4
- o Cut next to peak creates MACS problem, especially on older model, restricting combinations slightly.

3800 ICore (SFIC)

Examining the full line of KABA Peaks products requires a starting point, and the 3800 series is the logical starting point. On this model (and only this model, by the way) the Peaks patented pin is under the face cap, and is factory installed, even on uncombinated cores. Therefore, to pin the locks, it is merely necessary to select the appropriate standard SFIC pinning kit (A2, A3 or A4) and pin it as you would any other SFIC of that series, including the caps. You can use standard capping presses for it as well.

For A2 depths, you would use A series 0 to 9 Bottom Pins and B series 2 to 19 Top Pins.
For A3 depths, you would use C series 0 to 6 Bottom Pins and D series 1 to 13 Top Pins.
For A4 depths, you would use E series 0 to 5 Bottom Pins and F series 1 to 11 Top Pins.

(Remember that no new series should be created on the A3 cut depths.)

6800 ICore (SFIC)

On this model the Peaks patented pin is in the first position behind the face cap, and is field installed, except on combinated cores. Therefore, to pin the locks, it is merely necessary to select the appropriate standard SFIC pinning kit (A2, A3 or A4) and pin it as you would any other SFIC of that series after adding the patented pins.

Peaks obtained a new patent when the original expired by changing to a cupped bottom patented pin which resists wear or cutter errors.

For A2 depths, you would use A series 0 to 9 Bottom Pins and B series 2 to 19 Top Pins.
For A3 depths, you would use C series 0 to 6 Bottom Pins and D series 1 to 13 Top Pins.
For A4 depths, you would use E series 0 to 5 Bottom Pins and F series 1 to 11 Top Pins.
(Remember that no new series should be created on the A3 cut depths.)

The 6800 uses a pressed in place spring cover, similar to the brass slide covers on many mortise and rim cylinders on the market, but of stronger material to help resist "comb" attacks. Therefore using a special SFIC slide cover pressing block and the hand held slide cover tool, as shown in the following illustration, it is one simple operation to cover all the chambers at once. Therefore, if you will use .140" spacing ICores, it is faster and easier to use the 6800 than the 3800.

KSP

KSP   Product line has been around since before Patent ran out.
- Similar to Best core.
- Available in all "standard" keyways
- 5, 6 or 7 pin available .150 spacing
- Sold and stocked by many Locksmith Wholesalers

Lockwood

Lockwood   Product line has been around since Best Patent ran out.
- Tolerances in product line are good
- Similar to Best core but uses a brass slide cap instead of individual caps, but for the same purpose.
- Available in all "standard" keyways
- Manufactured in Australia
- 5, 6 or 7 pin available .150 spacing

Medeco Keymark (Also Yale)
(Original Sloped Leg Design)

Medeco Keymark (original design)

Uses specially designed key with sloped leg to resist duplication using standard equipment and limit access for picking.

Fig. 2

BLADE
Primary bitting area

LEDGE
- Reference and locating surface for key cutting on patented key punch.
- Available as warding or secondary bitting surface.
- Increased pick resistance

SECURITY LEG™
- Offset at 5-85° to produce an extraordinary number of unique keyways.
- Angled keyway resists pick and comb attacks.
- Prevents interchange with other keyways.

MILLING FOR KEYWAY WARDS
- Used to produce additional keyways within each security leg angle family.

CYLINDER PLUG SHEAR LINE
MASTER KEY (#3 DEPTH)
Fig. 4A

CYLINDER PLUG SHEAR LINE
CHANGE KEY (#5 DEPTH)
Fig. 4B

CONTROL SHEAR LINE (10 INCREMENTS HIGHER)
.125"(10 INCREMENTS)
CONTROL KEY (#6 DEPTH)
Fig. 4C

TOP PIN
7
11
2
3
TOTAL PIN STACK IS 23

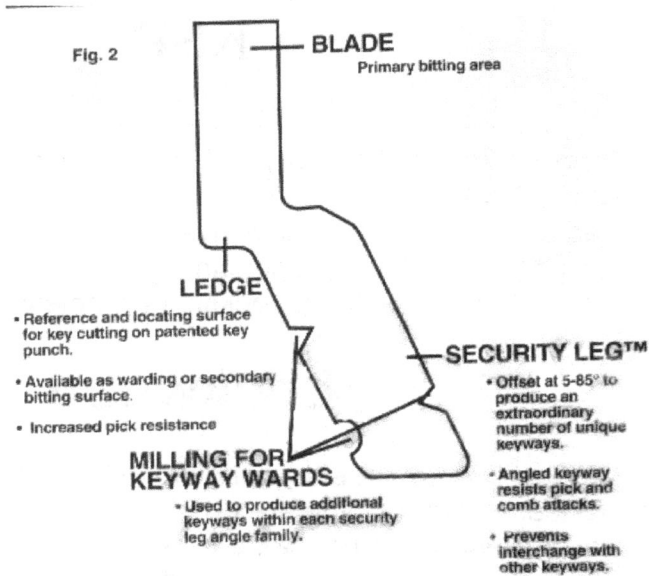

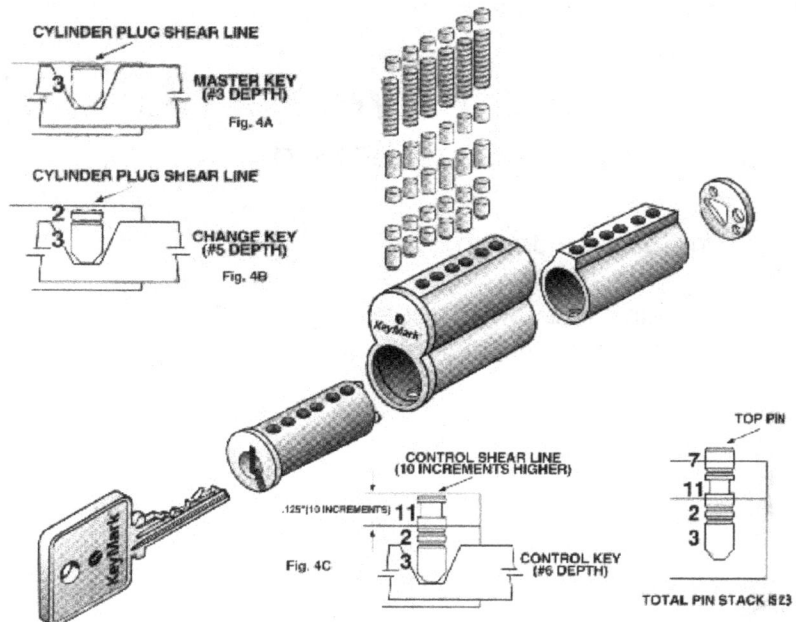

- Uses high quality nickel silver bottom pins to resist wear
- Uses "Spool Type" "mushroom" top pins to resist picking
- Some pinning kits and cores use a 19 stack height due to using a different spring length.

58

Medeco Keymark X4 (Also Yale)

Medeco X4

First generation KeyMark
with offset keyway

KeyMark X4

- o Special millings in the plug and control sleeve allow for a spring-loaded locking pin and slider that provide an additional locking point.
- o When in the locked position, the locking pin protrudes into the control sleeve (or shell in non-interchangeable core cylinders).
- o When the correct key is inserted, a special cut in the slider is moved into position allowing the pin to drop inside the circumference of the plug, thus allowing it to turn.
- o Since cylinder pinning is still accomplished through the top of the cylinder, the additional slider and pin components don't affect the ease of cylinder pinning.
- o offers higher security, improved key cutting capability and improved capabilities.

Sargent XC

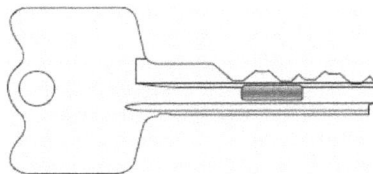

- o Similar to Best core but uses a saddle instead of a control sleeve, but for the same purpose.

- o Adds a "finger pin" assembly for increased security
- o Available in all "standard" (Best) keyways A through M and Q, as well as in Arrow's own 1C, 1D and restricted keyways
- o Plus Key can also work in ChoICe Base cores, permitting two levels of security in the same system or a phase-in approach.
- o Six pin and 7 pin available .150 spacing
- o Available in drill resistant version
- o Face plate is removable and replaceable to achieve the desired finish without carrying double the number of cores or more.
- o Furnished A2 only.

Schlage SFIC Classic

Schlage SFIC classic

- o Available in Best classic keyways A through M
- o .150 spacing only
- o Do NOT use Best original pins
- o Lab colored brass bottom pins available in universal pin shape for both Best and Schlage
- o High quality manufacturing

Schlage Everest B

- o Original Everest keys were protected by patents 5,715,715 and 5,809,816 until July 2014
- o Uses a check pin which is raised by a specially designed groove on the key.
- o B series not compatible with standard Schlage keyways
- o Not available in Best classic keyways

Scorpion CX-5

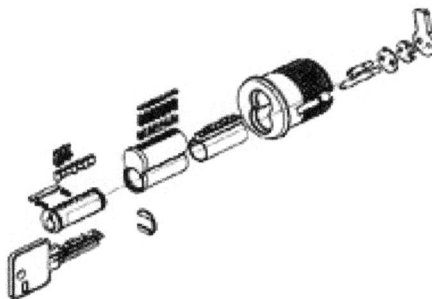

- o Scorpion CX-5 cores are available with press-in caps or screw type chamber retainers.
- o Keys have a wavy groove cut into the side which accommodates the legs of a sidebar.
- o Available in Locksmith only or assigned keyways

- o Excellent tolerances
- o Higher security feature without extra steps required to service it

Stanley SFIC

In 2011, Stanley took over the Best product line, and in 2013 they introduced their own value engineered, price-point version of the SFIC aimed at the commercial locksmithing market. Structurally it is similar to classic SFIC.

KABA Peaks SFIC and MFIC Products

Kaba Peaks was originally a Small Format Interchangeable Core product.

However, they later made special versions using the A2 and A4 depths from SFIC but having the outside shape of other manufacturer's LFIC products. Even though the control lug portion that is visible is the same as the original LFIC, however, the control sleeve covers all six positions. Each usually requires its own pinning kit because the plug diameter is larger than a standard SFIC.

Some consider this a hybrid between SFIC and LFIC and have named it Mixed Format Interchangeable Core (MFIC).

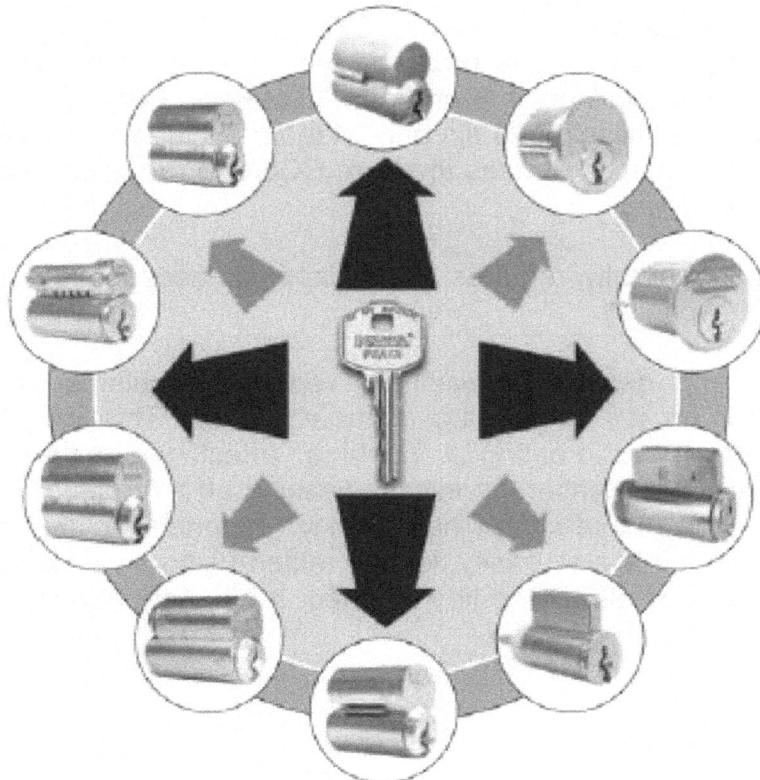

Originally introduced by Lori Lock as Lori Peaks, KABA now manufactures and sells these products as Kaba Peaks. They are also sold through Best distributors, who use different part numbers for the same product.

While not a high security UL437 type product, both the cylinders/cores and the key blanks are protected by utility patents with an expiration of the original patent in 2010, at which time a new patent was obtained for the Peaks Preferred line, which changed the patented peak pin to a cupped pin, resisting wear and cutter errors.

The name "Peaks" comes from a raised peak at the top of the blade of the key which contacts a "patented bottom pin" and raises it to the shearline. Lining up all the standard pins at the shearline accomplishes nothing unless the patent pin set is also lined up at the shearline. The key will not turn.

There are two patent bottom pins and two patent top pins. For small diameter plugs and SFIC, the patent bottom pin is part number 3800-00-3004 (Best part number PB1. For large diameter plugs, the patent bottom pin is part number 6140-00-3004 (Best part number PB2). The patent top pin for all core products (RCore and ICore) is 3800-00-3005 (Best part number PT1). The patent top pin for all non-core products (RCore and ICore) is 3425-00-3002 (Best part number PT2).

Peaks cylinders and cores are generally pinned like any other SFIC product, but there are some variations. The 3800/6800 series SFIC (Small Format Interchangeable Core) cores are .150" spacing, as are most of the SFIC on the market from various manufacturers, and are available in 6 or 7 pin. All of Kaba Peaks' other series, however are .140" spacing and available in 6 pin only. Kaba uses the term ICore for cores that fit into a standard SFIC opening. It uses the term RCore for any core that does not fit into the standard SFIC opening.

The .140" spacing generally requires hand capping, because the standard capping machines use .150" spacing.

Some locksmiths and dealers occasionally reported problems of cores not fitting properly, especially on the large plug diameter series. This was traced to excess chrome plating which was creating a "flashing" effect where the plug retainer was intended to fit. (Think of trimming the excess flashing off a model car or plane kit and it will make sense.) The manufacturer has quickly responded to any cases of this and corrected the problem. It is also easily field corrected using a wire wheel on the back of the plug with the retainer off and no key inserted. It has not been reported on the SFIC product.

Pin Sizes (approximate):

Depths	A2 BP	Depths	A2 TP	Depths	A3 BP	Depths	A3 TP	Depths	A4 BP	Depths	A4 TP
0A	.110"			0C	.110"			0E	.110"		
1A	.122"			1C	.128"	1D	.018"	1E	.131"	1F	.021"
2A	.135"	2B	.025"	2C	.146"	2D	.036"	2E	.152"	2F	.042"
3A	.147"	3B	.037"	3C	.164"	3D	.054"	3E 0K	.173"	3F	.063"
4A	.160"	4B	.050"	4C	.182"	4D	.072"	4E 1K	.194"	4F	.084"
5A 0J	.172"	5B	.062"	5C	.200"	5D	.090"	5E 2K	.215"	5F	.105"
6A 1J	.185"	6B	.075"	6C	.218"	6D	.108"	3K	.236"	6F	.126"
7A 2J	.197"	7B	.087"			7D	.126"	4K	.247"	7F	.147"
8A 3J	.210"	8B	.100"			8D	.144"	5K	.268"	8F	.168"
9A 4J	.222"	9B	.112"			9D	.162"			9F	.189"
5J	.235"	10B	.125"			10D	.180"			10F	.210"
6J	.247"	11B	.137"			11D	.198"			11F	.231"
7J	.260"	12B	.150"			12D	.216"				
8J	.272"	13B	.162"			13D	.234"				
9J	.285"	14B	.175"								
		15B	.187"								
		16B	.200"								
		17B	.212"								
		18B	.225"								
		19B	.237"								

Spool bottom pins to resist picking are available in 7A/2J, 8A/3J,and 9A/4J for the A2 system. The appropriate number is 7AS, 2JS, 8AS, 3JS, 9AS, 4JS respectively.

Spool top pins to resist picking are available in 6BS, 8BS,and 10BS for the A2 system. Spool bottom pins to resist picking are available in 4E/1K, 5E/2K for the A4 system. The appropriate number is 4ES, 1KS, 5ES, 2KS respectively.

Spool top pins to resist picking are available in 4FS, 5FS,and 6FS for the A4 system. Let's examine each series:

3800 ICore (SFIC)

Examining the full line of KABA Peaks products requires a starting point, and the 3800 series is the logical starting point. On this model (and only this model, by the way) the Peaks patented pin is under the face cap, and is factory installed, even on uncombinated cores. Therefore, to pin the locks, it is merely necessary to select the appropriate standard SFIC pinning kit (A2, A3 or A4) and pin it as you would any other SFIC of that series, including the caps. You can use standard capping presses for it as well.

For A2 depths, you would use A series 0 to 9 Bottom Pins and B series 2 to 19 Top Pins.
For A3 depths, you would use C series 0 to 6 Bottom Pins and D series 1 to 13 Top Pins.
For A4 depths, you would use E series 0 to 5 Bottom Pins and F series 1 to 11 Top Pins.

(Remember that no new series should be created on the A3 cut depths.)

6800 ICore (SFIC)

On this model the Peaks patented pin is in the first position behind the face cap, and is field installed, except on combinated cores. Therefore, to pin the locks, it is merely necessary to select the appropriate standard SFIC pinning kit (A2, A3 or A4) and pin it as you would any other SFIC of that series after adding the patented pins.

For A2 depths, you would use A series 0 to 9 Bottom Pins and B series 2 to 19 Top Pins.
For A3 depths, you would use C series 0 to 6 Bottom Pins and D series 1 to 13 Top Pins.
For A4 depths, you would use E series 0 to 5 Bottom Pins and F series 1 to 11 Top Pins.
(Remember that no new series should be created on the A3 cut depths.)

The 6800 uses a pressed in place spring cover, similar to the brass slide covers on many mortise and rim cylinders on the market, but of stronger material to help resist "comb" attacks.

Therefore using a special SFIC slide cover pressing block and the hand held slide cover tool, it is one simple operation to cover all the chambers at once. Therefore, if you will use .140" spacing ICores, it is faster and easier to use the 6800 than the 3800.

Standard Mortise and Rim Cylinders

Although the SFIC cores can be used in a mortise or rim housing, KABA Peaks also offers a standard mortise cylinder in 1-1/8 and 1-1/4 inch lengths as well as a 5/32" ring to shorten the 1-1/8 to accommodate Adams-Rite and similar locksets. The mortise cylinder is the 3401 and the rim cylinder is the 3402. Both use the standard pinning that the SFIC for that series of cut depths would use, although the control key will not operate the cylinder, so the control pins could be eliminated and a longer driver used if desired.

For A2 depths, you would use A series 0 to 9 Bottom Pins and B series 2 to 19 Top Pins.
For A3 depths, you would use C series 0 to 6 Bottom Pins and D series 1 to 13 Top Pins.
For A4 depths, you would use E series 0 to 5 Bottom Pins and F series 1 to 11 Top Pins.

(Remember that no new series should be created on the A3 cut depths.)

6240 RCore and 6340 RCore

KABA Peaks did not stop with the standard SFIC and the mortise and rim cylinders. In a move that was then unprecedented, they also made the 6240 RCore to fit the Yale Removable Core opening and the 6340 to fit the Medeco standard 6300 series Removable core opening. In both cases, the control sleeve covers the chambers in all six positions, and the locks are pinned exactly as they would be for a standard SFIC core. The only difference is the addition of the patented bottom and top pin just behind the face of the core.

For A2 depths, you would use A series 0 to 9 Bottom Pins and B series 2 to 19 Top Pins.

For A3 depths, you would use C series 0 to 6 Bottom Pins and D series 1 to 13 Top Pins.
For A4 depths, you would use E series 0 to 5 Bottom Pins and F series 1 to 11 Top Pins.

(Remember that no new series should be created on the A3 cut depths.)

6140 RCore and 6540 RCore

Not all removable core products were adaptable to the small diameter plugs, so KABA Peaks also made large plug diameter removable cores. These cannot be pinned using the standard SFIC bottom pins because the plug diameter is approximately .0625" larger, requiring each bottom pin to be proportionately higher. KABA Peaks provides the longer pins to fit the A2 and A4 series cut depths only, meaning that they cannot be pinned to the A3. The longer bottom pins to fit the A2 cut depths is called J series and the longer bottom pins to fit the A4 is the K series.

The large plug diameter RCores are:

      6140 to fit Corbin-Russwin Removable Core
      6540 to fit Sargent Removable Core

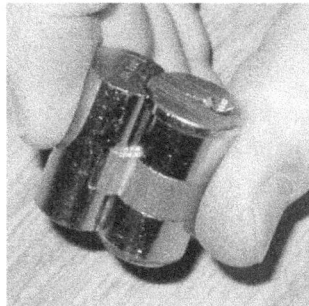

The 6540 is a direct replacement for this Sargent 6300 shown here.

Because the plug diameter is .0625" larger than the SFIC plug diameter, the bottom pins for each are also proportionately larger than the standard SFIC bottom pins.

In fact, for the A2 cut depths, a number 5A pin is the same as a OJ pin. A 6A is the same as a 1J pin. The 7A pin is the same as a 2J pin. The 8A pin is the same as a 3J pin. The 9A pin is the same as a 4J pin.

Similarly, for the A4 cut depths, a number 3E pin is the same as a OK pin. A 4E is the same as a 1K pin. The 5E pin is the same as a 2K pin.

Notice that the increment size did not change, only the actual pin length. For that reason, the top pins do NOT change at all. For the A or J series (A2) you still use the B series top pins. For the E or K series (A4) you still use the F series top pins.

Because the A2 bottom pins are still numbered 0 to 9 and the A4 bottom pins are still numbered 0 to 5, the pinning calculations remain the same as they would be for the SFIC in the A2 or A4 respectively. The stack height remains the same.

For A2 depths, you would use J series 0 to 9 Bottom Pins and B series 2 to 19 Top Pins.

For A4 depths, you would use K series 0 to 5 Bottom Pins and F series 1 to 11 Top Pins.

6440 RCore to fit Full Size Schlage Core Openings

KABA Peaks also makes an RCore to fit the full size Schlage removable core housings (which Schlage refers to as FSIC or Full Size Interchangeable Core) but due to the size of the original core for that, special pins had to be used for both the bottom and top pins for the KABA Peaks version, which is the 6440 series RCore. These use bottom pins of 0X to 9X and top pins of 2W to 19W and can only be pinned to the A2 series, not the A3 or A4 because the corresponding pin sizes are not manufactured. Once again the control sleeve covers all the six chamber positions, but the control lug remains in the 7th, requiring a special control key blank to install or remove these cores. Pinning is exactly like any other A2 series SFIC, except that it requires a special pinning kit.

For A2 depths, you would use X series 0 to 9 Bottom Pins and W series 2 to 19 Top Pins.

3400 series to fit key in knob and lever locksets

Although the focus of this book is interchangeable and removable cores, it should be noted that KABA Peaks also makes a series of cylinders to fit most manufacturer's key in knob or key in leversets, as well as some padlocks, allowing these to be keyed to the same key system as the RCores.

This is the 3400 series and there are two models, one with a C-clip on the back of the plug as a plug retainer, which uses the small plug diameter bottom pins (A,C,E), and one with a screw cap for a plug retainer, which uses large plug diameter pins (J, ,K).

It should be noted that some of each of these may require special tailpieces or adaptor kits to fit some manufacturer's products. Some of these are easily recognized and will be noted here, but there are so many products to fit that many will not be listed here. Kaba has an Access Control Catalog available that includes the KABA Peaks in more detail than this book is intended to achieve. It is an excellent source of information, especially on all the kits and adaptors and tailpieces.

3400 Series Small Plug Diameter

Because the 3400 series fits so many brands of locksets, most of which use the large plug diameter and longer bottom pins, it is easier to begin by identifying which models use the standard small plug diameter (A,C,E) bottom pins.

They are:

3400-xx-1004	Corbin-Russwin CK4200 Grade 1 Knob Lock
3400-xx-1004	Corbin-Russwin UT5200 series Unit Lockset
3400-xx-1055	Corbin-Russwin CL3900 series Leverset
3400-xx-1008	Sargent 7, 8, and 9 grade 1 Knob Lockset
3400-xx-1076	Sargent 7600 Integralock
3400-xx-1054	Yale 5300 Grade 2 Knob Lockset
3400-xx-1054	Yale 5400 Grade 1 Knob Lockset
3400-xx-1055	Yale 540F Trim

Using the 05 adaptor kit, KABA Peaks small plug diameter cores can fit the following:

3400-xx-1005	Sargent 11 Line T Zone Leverset

Using the 06 adaptor kit, KABA Peaks small plug diameter cores can fit the following:

3400-xx-1001	American Padlock 3600 and 3700 series
3400-xx-26	Arrow M series Tudor and Darrin Knob Locksets
3400-xx-01	Master System 29 Padlocks
3400-xx-03	Trilogy 2700 and 3000 series Leversets

Using the 06 cylinder adaptor kit and field modifying the tailpiece, KABA Peaks small plug diameter cores can fit the following:

Corbin-Russwin 3391 CRM Classroom Crisis Leverset

3400-xx-1095 and 3400-xx-1099 Large Plug Diameter

The remainder of the KABA Peaks 3400 series cylinders are the -1095 and -1099 series, using a large plug diameter and the longer (J and K) series bottom pins. These use a screw cap for a plug retainer. Remember that the top pins are still B series for A2 and F series for A4.

Review/Overview of KABA Peaks Products

Pins:	Product Line
A,B (A2)	**3800** SFIC with chamber caps
C,D (A3)*	**6800** SFIC with spring cover
E,F (A4)	**3401** Mortise cylinder (1- ¼" , 1- 1/8" and 5/32" ring)
	3402 Rim Cylinder
	6240 to fit Yale Removable Core
	6340 to fit Medeco 6300 series
	3400 small diameter cylinder (various- with C-clip plug retainer)
	3400-xx-1004 Corbin-Russwin CK4200 Grade 1 Knob Lockset
	3400-xx-1004 Corbin-Russwin UT5200 series Unit Lockset
	3400-xx-1055 Corbin-Russwin CL3900 series Leverset
	3400-xx-1008 Sargent 7, 8, and 9 grade 1 Knob Lockset
	3400-xx-1076 Sargent 7600 Integralock
	3400-xx-1054 Yale 5300 Grade 2 Knob Lockset
	3400-xx-1054 Yale 5400 Grade 1 Knob Lockset
	3400-xx-1055 Yale 540F Trim
	Using the 05 adaptor kit, KABA Peaks small plug diameter cores can fit:
	3400-xx-1005 Sargent 11 Line T Zone Leverset
	Using the 06 adaptor kit, KABA Peaks small plug diameter cores can fit:
	3400-xx-1001 American Padlock 3600 and 3700 series
	3400-xx-26 Arrow M series Tudor and Darrin Knob Locksets
	3400-xx-01 Master System 29 Padlocks
	3400-xx-03 Trilogy 2700 and 3000 series Leversets
J,B (A2)	6140 to fit Corbin-Russwin Removable Core
K,F (A4)	6540 to fit Sargent Removable Core
	3400 large diameter (-1095,-1099)- (various-with screw cap)
X,W (A2)	6440 to fit Schlage Full Size (FSIC)

* No new systems should be created on A3 due to system flaws

Chapter Seven:
Tools and Techniques

The right tools make the job a LOT easier. If the spring retainer cap is the pressed in brass slide type, you will need an ice pick or similar tool to pry up the cap so that you can remove the old pins and re-pin through the top, as shown in the next illustration.

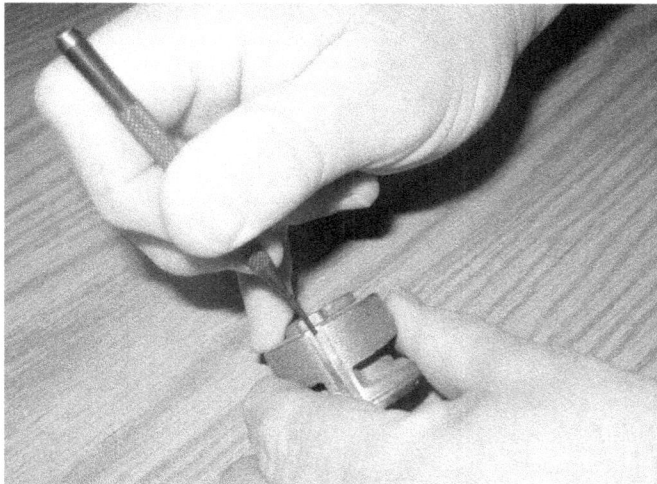

Fig 7-1) An Ice pick , hook or similar tool is used to pry the spring retainer cap up

When it comes time to put the new spring retainer cap back on, the first thing you will need is the new brass cap. Then you will need something to hold the cylinder in, such as the staking fixture shown in fig 7-2

73

Fig 21-2) A staking fixture. This one is for Corbin Cores but they all work in a similar manner.

And finally, you will need a staking tool. Some locksmith try to get by with an old dull screwdriver or a punch of the appropriate size, and these will work if you are very careful, but it is better to pick up the proper tool. Not only does it make the job much easier and faster, but it also seats the slide cap more positively without creating weak spots. One such tool is shown in the next illustration. These tools are a bit pricey, but well worth it if you do a lot of this sort of work.

fig 7-3) A staking tool in action, about to be struck by a hammer.

Of course, if you absolutely MUST make your own staking tool, a ¾ inch cold chisel can be ground and filed to the correct dimensions, and will not damage the core the way a screwdriver tends to.

Pinning Kits

No matter which manufacturer shoes cores you choose to work on, you will need a pinning kit. Lab makes pinning kits for each manufacturer, and this is one place not to save money. Get all the kits you will need, because there is no easy alternative.

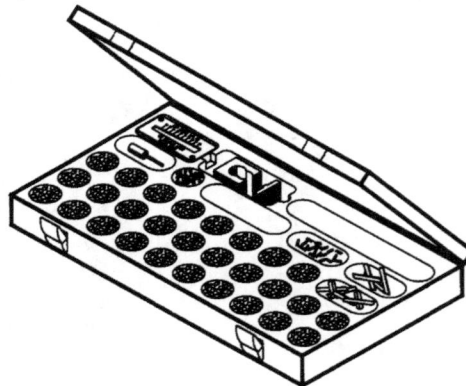

74

Remember that when it comes to SFIC there are five basic kits...A2 for Best which has bottom pins with a small flat, A2 for Schlage or Falcon, which tend to fit most other brands except Best, the colored brass Universal SFIC A2 pin kit, the A3 kit and the A4 kit. If you are using A4 on anything but Best add another kit for that. To save you a little money, Lab makes a combination A3/A4 kit.

Peaks bottom pins are .003" shorter than the other brands, and Kaba does not recommend using other brands of pins in their cores with .140" spacing.

You will find the staking fixtures or core holding fixtures or capping/disassembly blocks useful or any products you are working with. Lab, Major, A1 and Arrow all have excellent ones for some of the products, but a sheet of thick plastic and a small drill assortment will allow you to make excellent ones if you need to save money, because they can be a bit pricey to purchase, particularly if you service a wide line of products.

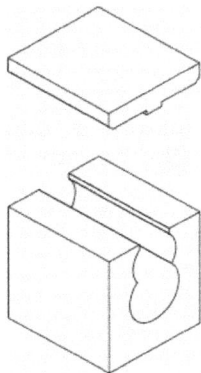

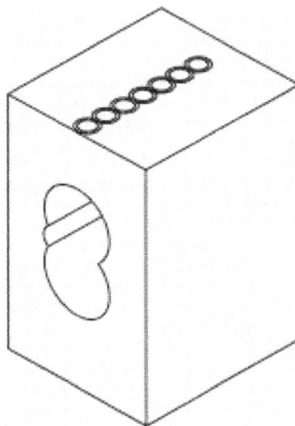

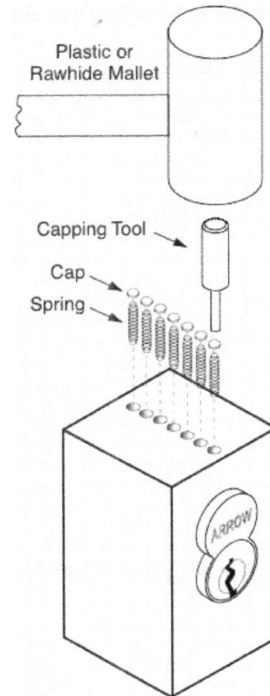

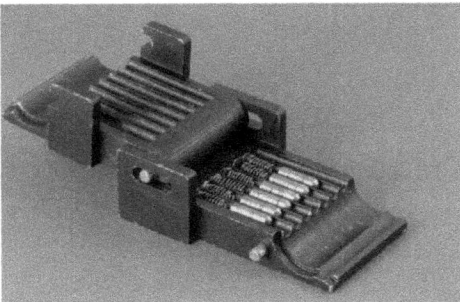

Decoding Block

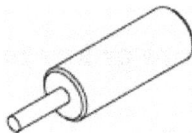

Only a plastic or rawhide mallet should be used for small format capping.

Stamping equipment can be manual or automatic, depending upon your needs.

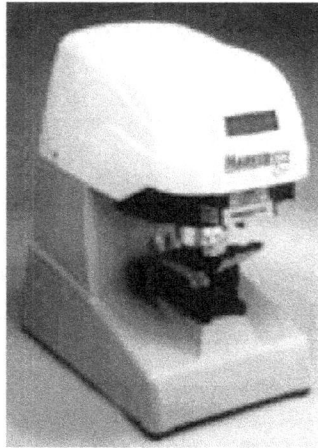

CAPPING BLOCKS differ slightly from standard holding fixtures although there are some that are both. A capping block ins intended to hold the combinated core while installing springs and caps.

EJECTOR TOOL

Ejector Tool

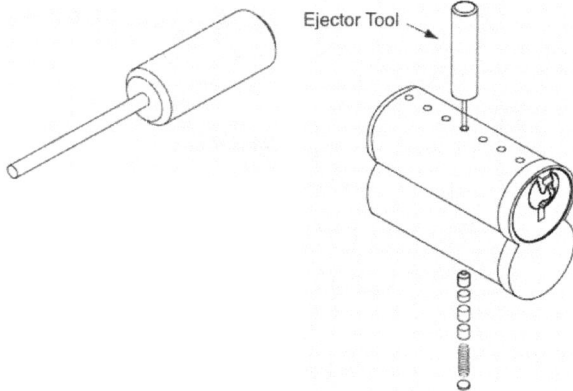

This is used on SFIC that feature a capped chamber and an access hole in the bottom of the core, which not all SFIC do.

Stamping Plates can be for one core or key or for up to 20.

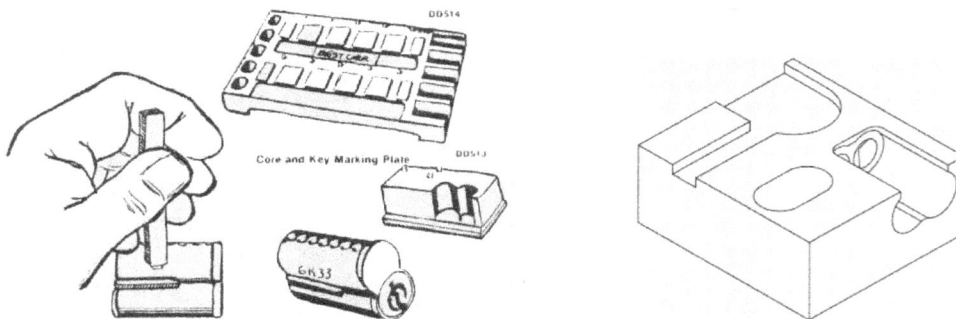

Core and Key Marking Plate

CYLINDER INSTALLATION WRENCH

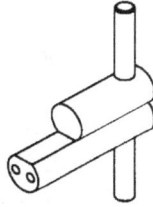

These are useful, and are available for Sargent and Corbin Russwin LFIC models, as well as for SFIC

KEY GAUGEs are available for almost any product.

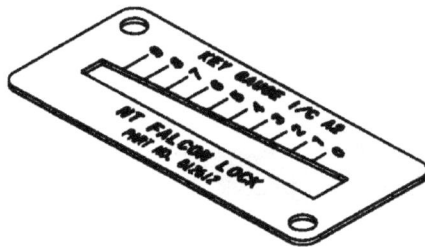

CAPPING PRESS
If you service large quantities of cores with individual caps, a capping press is the most efficient piece of equipment.

The capping press is just an arbor press with a capping block attached to it. Capping presses also have a small tray to hold caps for quick insertion.

When all the pins are in the core, insert the core into the capping block of the press. Insert a spring into each chamber and then slide caps from the tray into each pin chamber.

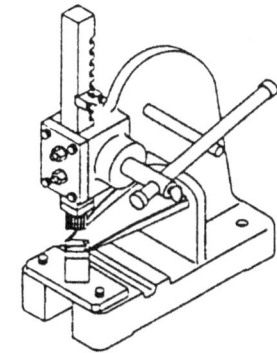

Pulling down on the handle forces a block with six or seven "fingers" down into the capping block, sealing all chambers of the core in a single operation.

A-1 Security Mfg. designed the **Capsaver**, which punches the caps out of strips of brass and seals all chambers of the core in a single motion. Each strip can be inserted in four different orientations in a slot in the top of the capping block, allowing you to cap four cores out of one strip of brass. It is quick and efficient but not for rekeying a core where a single position had to be serviced.

KEY PUNCH

Called a "key combinator" by Best, punch type machines are very popular for SFIC work.

Best originally color-coded their punches. Red was A2, Yellow-green was A3 and orange was A4. Later they went to a modifiable brown machine that could be set up by the factory for whichever was needed for a particular customer.

The Mean Green Machine and the Pak-A-Punch by A-1 Security Mfg. are other fine products for punching IC keys. The Mean Green Machine is patterned after the old style Best combinatory, but with a hand-friendly wider handle.

The Pak-A-Punch is a portable hand-held punch if you need the option of generating keys on-site. It is not especially suited to large jobs, however.

Pro Lock makes a blue colored field changeable punch using depth plates and replaceable vise jaws, making it the one of the most versatile of the group.

A1 Green

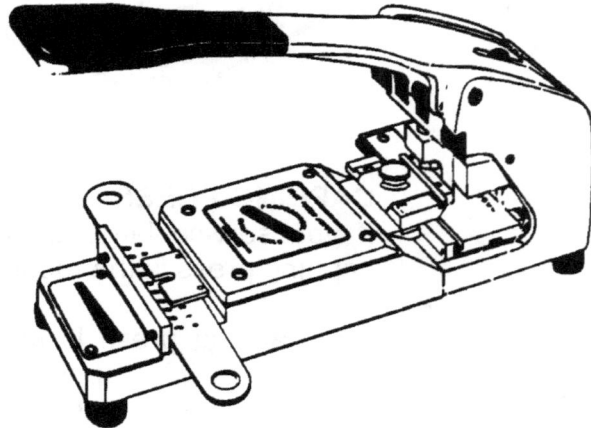

Pro-Lock Blue Punch

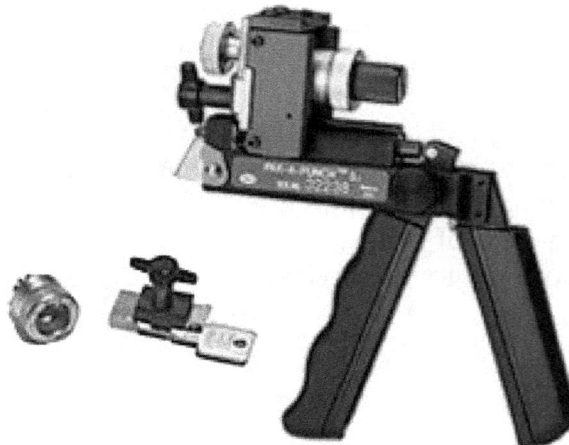

A1 Pak-a-Punch

A word about SFIC pins

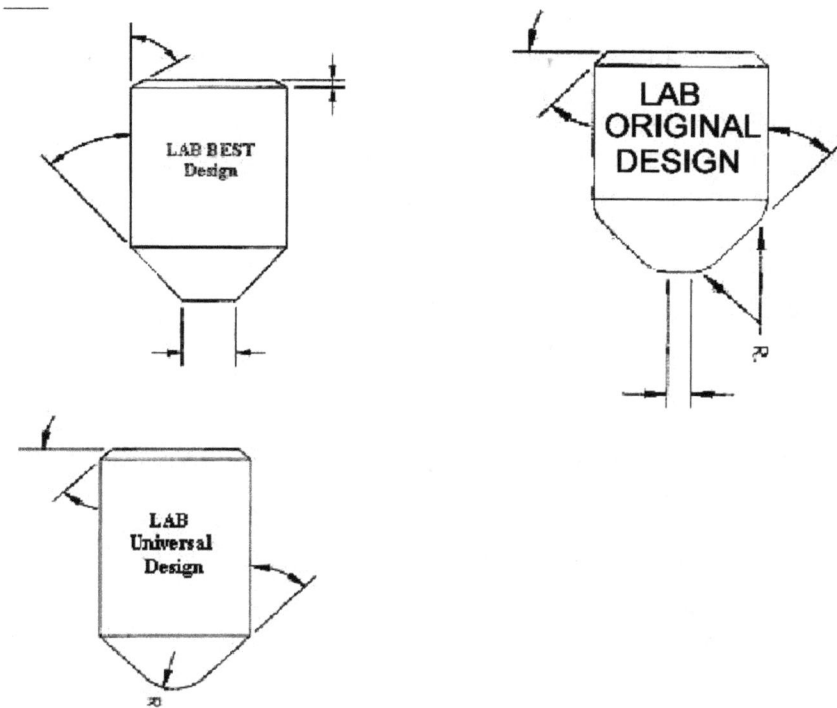

There are three type of pins made for SFIC products.

The flat tipped ones as shown on the left above are intended for Best and Peaks only.

The rounded pins shown in the center above fit most other manufacturers of SFIC.

The universal pins, shown on the right, fit all of these but are only available in brass pins – not nickel silver as the manufacturer all suggest, and are color coded which, while handy, can be considered a security risk as well.

Choose your pin kits carefully. It is better to have a kit you rarely need than not have a kit that works reliably. Pin kits are relatively inexpensive, take up very little room, and are genuine time savers, so having extra kits is generally not a bad choice.

There are no specific kits available for the Mixed Format cores at the time of this publication, but Lab sells empty customizable kits also.

Core Extraction

It is not uncommon to find a business has moved and left its cores in place at the old location with no contact information being passed on to potential new occupants.

The core manufacturers will generally not make the control key and master key information available to the new occupants because to do so may compromise any systems taken by the original owner to their new location, opening up serious litigation potential. Similarly, neither should the Locksmith.

However, cores are far less expensive than most of the institutional locksets, so there may be a genuine, and justified, desire to keep the hardware and simply replace the cores.

This means the Locksmith will need to remove the existing cores, preferably with no damage to the locksets or tailpiece actuators.

One way to open the locks is to pick the cores. However, it is difficult to pick an interchangeable core because there is a tendency for some pins to line up at each shearline, and as you already know, this accomplishes nothing. Not every core manufacturer has chamber access holes. Best does, however, and many others do as well. If the core has chamber access holes, a specially designed tension wrench (Available pre-made through distributors or which can be hand made in a pinch by filing an existing tension wrench) can apply turning pressure only to the upper (control) shearline.

To do this, the tension wrench is ground down to permit one or more tips approximately .015 inches to .020 inches in length, facing the bottom of the keyway. These grab the access holes in the control sleeve, and apply tension to the control sleeve instead of to the plug directly. Therefore, when the core is picked , it will be to the removal position.

Once the core is out in your hands, the spring cover can be removed, and the drivers measured. Remember that the driver pin sizes are a constant for any given customer's control key, so one core allows you to create a control key to remove all the existing cores, either for re-pinning or (recommended) for replacement.

The chart on the following page shows the driver lengths corresponding to the control cuts:

Driver Length to Control Key Cut Conversion

A2

.050"	9
.062"	8
.075"	7
.087"	6
.100"	5
.112"	4
.125"	3
.137"	2
.150"	1
.162"	0

A3

.054"	6
.072"	5
.090"	4
.108"	3
.126"	2
.144"	1
.162"	0
.054"	6
.072"	5
.090"	4

A4

.063"	5
.084"	4
.105"	3
.126"	2
.147"	1
.168"	0

Drilling Cores

Drilling should be considered a last resort method of core extraction. With the exception of a few high security products, it is not all that difficult, but it is easy to damage the pins beyond the capability of being decoded, or worse yet, to damage the lockset some of which sell for several hundred dollars each.

There are several techniques available for drilling. Some Locksmiths prefer to damage one lockset knob or lever (they often have a replacement available from an older take-out) than risk not being able to decode the core. In this case, they drill the lug off the core, or the housing in front of the lug. This is done by drilling just to the left of the center of the figure 8.

Others prefer to insert a broken off blade from a blank, raising the tumblers above the shearline, and drill at the top edge of the plug directly above the keyway. The plug is then rotated, and the upper pins removed with a tweezer. This is more difficult than it sounds.

Still others prefer to drill at the upper shearline until they are past the face of the plug, and can see the control lug. Then they apply direct pressure to the lug and pick the lock. This is done by drilling approximately 1/8 inch above the top of the plug.

Arrow cores, with a removable face plate, are often extracted using this method.

There are also jigs available which include a tool to raise the pins, then drill the core, turn the control sleeve, and decode the core's driver pins which are hopefully undamaged.

And of course, if decoding seems like more work than drilling (which I cannot personally see, but have heard occasionally) every core can be removed by drilling 1/8 inch above the plug in line with the keyway, and drilling through the number of pins in the core, then turning the control sleeve with a screwdriver.

The worst job of drilling I ever saw was where a Locksmith had used a ½ inch bit to remove the plugs completely. But it did save the locksets, though they all needed cleaned out and lubricated before being re-used.

Pinning Calculators

There are a wide variety of small programs and apps on the market for calculating the pins for SFIC products. Blackhawk Software seems to be the fore-runner of it, but it seems to be the first app each rising would-be programmer/data technician writes, so there is an abundance of them on the market, including simple ones written for Spreadsheet programs. These are all inexpensive, but people seem to constantly expect more from them than they were intended to do. They do NOT lay out a system, for example. But most do what is promised-calculate the pins. After all, it is a simple process:

To calculate the bottom pins, calculate the smaller of the two cuts for each position between the Master Key and Individual Key (Change Key).
example from dBase3:

BP1 = Min(Mcut1,Icut1)

example from spreadsheet where C10=CTKC1, C8=CKC1, C7=MKC1, C5=BP1, C4=MP1, C3=CP1, C2=DR1:

C5=MIN(C8,C7)

To calculate the master pins, calculate the absolute value of the difference between the two cuts for each position between the Master Key and Individual Key (Change Key).

example from dBase3:

MP1=ABS(Mcut1-Icut1)

example from spreadsheet where C10=CTKC1, C8=CKC1, C7=MKC1, C5=BP1, C4=MP1, C3=CP1, C2=DR1:

C4=ABS(C7-C8)

To calculate the Control Reference Number, calculate the control additive plus the control key cut for each position:

example from dBase3 for A2:

CRN1=Ccut1+10

example from spreadsheet where C10=CTKC1, C8=CKC1, C7=MKC1, C5=BP1, C4=MP1, C3=CP1, C2=DR1:

C11=+C10+10

To calculate the Driver Pin, calculate the Stack Height minus the Control Reference Number for each position or calculate the Stack Height minus the sum of Bottom Pin, Master Pin and Control Pin.

examples from dBase3 for A2:

Dr1=23-CRN1
or
Dr1=23-(BP1+MP1+CP1)

example from spreadsheet where C10=CTKC1, C8=CKC1, C7=MKC1, C5=BP1, C4=MP1, C3=CP1, C2=DR1:

C2=23-(C11)
or
C2=23-(C5+C4+C3)

To calculate the Control Pin, calculate the Control Reference Number for each position minus the largest cut between the two cuts for each position between the Master Key and Individual Key (Change Key).

example from dBase3:

CP1=(CRN1-(Max(Mcut1,Icut1))

example from spreadsheet where C10=CTKC1, C8=CKC1, C7=MKC1, C5=BP1, C4=MP1, C3=CP1, C2=DR1:

C3=(+C10+10)-(C5+C4)

The actual verbiage varies with different programming languages, but the math routines are the same.

You can also use a speed chart like the ones shown in this book, or a cube such as that sold by Xpertinex:

http://www.iclsglobal.com/QU-I.C-Key.html

System Layout

Creating a key system for SFIC products is no more difficult than for any other type of lock cylinder. Most manufacturers use Total Position Progression. A2 products use two increment drop, and most have no MACS violations. A3 was originally single increment, then modified to a limited single increment, keeping a two increment step between master cuts and change key cuts and a single increment between change keys. Most systems remaining on A3 use two increment. A4 use single increment.

Although you lose no keys to MACS violations, however, you do lose some to whatever rule is in play regarding control key selection and preventing change keys from operating as a control key and removing cores. Some system layout personnel simply assign one bitting from one or two positions of the KBA and reserve it exclusively to the control key, wiping out ALL combinations using it. Others choose a bitting and arbitrarily cross off bittings they feel are too close to it, so that not as many bittings are lost.

For the A2, a popular technique was to use a nine cut in an even parity position and then not use an eight cut on any change keys in that position.

If any of the terms in this section are unfamiliar to you, we recommend the following book:

Master Keying Textbook by Don OShall and Tony West

Notes: